CONTENTS

Ceramics:
applications and limitations

M H LEWIS

Within the past decade a number of
significant advances have been made in
the development of new ceramics which
has been stimulated by the inadequacy of
metallic alloys or traditional ceramics in
meeting the needs of newly-developing
technologies. This paper surveys some
of the current and potential applications
of the new ceramics in relation to their
microstructure and properties. The main
emphasis is on monolithic ceramics based
on Si_3N_4 and SiC, which have important
applications in high-temperature energy
conversion systems, and on the ZrO_2-
toughened ceramics, currently restricted
to a lower temperature regime. Ceramic/
ceramic composites are briefly reviewed
as an important new class of material
which may overcome some of the
engineering design/reliability problems
associated with brittle fracture.

The author is with the Centre for
Advanced Materials Technology,
Department of Physics, University
of Warwick.

INTRODUCTION

The name 'ceramic' has traditionally been
associated with the chemically and
microstructurally complex products of 'firing'
natural silicate minerals following a
convenient shaping operation in which the
silicate particles are held in a fluid medium.
A large volume of the ceramic industry, based
on the need for cheap sources of thermal or
electrical insulation, will continue in this
'traditional' ceramic area. However, there has
been an increasing requirement for operation of
ceramics in an environment of greater severity
which has stimulated the development of
high-purity synthetic ceramics with enhanced
chemical, thermal or mechanical properties.
These are the oxide refractories, based mainly
on Al_2O_3 MgO or ZrO_2, and more recently the
nitrides or carbides of silicon and boron and
the silicides or borides of selected transition
elements (e.g. $MoSi_2$, TiB_2).

In the 1980's we are witnessing a new era
in the development of structural ceramics based
principally on the evolution of microstructures
with special mechanical or thermo-mechanical
properties. A well-publicised example of a
recent 'high-technology' ceramic application is
that of the thermal protection system of the
space shuttle. The requirement for low density
and thermal conductivity, high thermal-shock
resistance together with a retention of
moderate strength and rigidity has been met by
a combination of intrinsic solid properties and
microstructural design. Thus the insulating
layer, which has to withstand sudden re-entry
temperatures up to 1400°C and large 'g' forces,
consists of micron-diameter SiO_2 glass fibres
in a random array, leaving >90% porosity
(Fig.1). The necessary intrinsic properties
are conferred by the disordered (glassy) SiO_2
structure which has 1/10 of the thermal
conductivity of crystalline SiO_2 and an
extremely small thermal expansion coefficient
($0.5x10^6$ compared e.g. to Al_2O_3 - $8.8x10^6$). The
low density - high porosity microstructure
enables a further large reduction in thermal
conductivity and rigidity is provided by
high-temperature diffusional bonding between
the glass fibres.[1]

This introductory example of the
evolution of a ceramic with appropriate
macroscopic properties is a simple illustration
of the interaction between intrinsic
phase-properties and microstructural
'engineering'. Similar principles apply to the
wide variety of phase combinations and
morphologies found in the more complex
microstructures of the new 'engineering

Table 1

Estimated energy gain in using ceramic components in heat engines. (Ref.2)

Type of engine	% Reduction in fuel
Automotive turbine 150 h.p., 1370°C inlet temperature.	27
Truck diesel (adiabatic turbo-compound) 500 h.p., 1210°C max. component temperature.	22
Passenger car diesel (no cooling) 80 h.p., 800°C max. component temperature.	10-15

Table II – Si_3N_4 – based ceramics

Major phase : normally βSi_3N_4 or substituted-derivative $\beta'(Si_{3-x}Al_xO_xN_{4-x})$		
Type	**Example of Composition and Source**	**Microstructure**
HPSN or HIPSN	1% MgO additive $\qquad$ Norton NC132 2-5% Y_2O_3 additive $\qquad$ Battelle–U.S.A.	0-3% mainly glassy intergranular residue
SSN	5% MgO + 9% Al_2O_3 $\qquad$ Kyocera SN205 6% Y_2O_3 + variable Al_2O_3/AlN $\qquad$ Lucas–Cookson–Syalon 101/201 6% Y_2O_3 $\qquad$ GTE PY6	5-10% glassy or crystalline matrix + β' βSi_3N_4 + glass
RBSN	Nitriding of Si during sintering without additives $\qquad$ Norton NC350 AED 'Nitrasil'	60-90% αSi_3N_4 10-40% βSi_3N_4 5% Si 15-30% porosity
SRBSN	Similar to RBSN, followed by high-temperature sinter (with additives)	Similar to SSN

Table III – SiC-based Ceramics

	Major phase : either αSiC (hexagonal polytypic) or βSiC (cubic)		
Type	Example of Composition and Source		Microstructure
HP SiC	1% B_4C or Al/Al_2O_3	Norton NC203 Elektroschmelzmer Co. FRG	αSiC + up to 3% intergranular residue with Al/Al_2O_3 additions
SSiC	0.5–1% B + C or Al+C additives	Kyocera SC201 Carborundum Co. G.E. 'Sintride'	" " βSiC
SiSiC	'Siliconized' : wide range of composition and sintering process e.g. SiC + graphite preform + reaction with Si in liquid or vapour state	Norton NC435 Coors	α + β SiC + ∿10% free Si

Table IV – ZrO_2-toughened ceramics

Type	Example of Composition and Source	Microstructure
TZP	2–4% Y_2O_3/ZrO_2 NGK Z191 sintered 1350–1500°C	Fine-grained, fully tetragonal polycrystal
PSZ	8–10% MgO/ZrO_2 Feldmuhle ZN40 } sintered 1650–1850°C, Coors } heat–treated 1100–1450°C }	Tetragonal ZrO_2 precipitates in cubic matrix

Property Comparison

Type	MOR (MPa)	K_{1c} (MPa m$^{1/2}$)
TZP	1000–2500	7–12
PSZ	600– 800	6– 8
ZT(Al_2O_3)	500–1300	5– 8
CSZ (fully cubic)	200	3
SSN	500– 800	5– 8
SSiC	400– 600	3– 5

ceramics' discussed in this paper. The emphasis will be on structural ceramics in which the development of exceptional mechanical or thermo-mechanical properties is of prime concern.

2. THE PROCESSING CONSTRAINT

The idealised sequence of phase selection and microstructural engineering, outlined above, is frequently inhibited via kinetics or thermodynamics (phase equilibria) associated with the ceramic fabrication process. A melting/casting/thermal-mechanical process commonly used in metallurgical practice is inappropriate for ceramics in view of their high melting temperatures or decompositional problems, susceptibility to thermal shock and resistance to plasticity. The nearest approach is that of silicate glass melting, shaping and controlled crystallisation typified by the 'glass ceramic' processing route. More frequently ceramics are formed by the consolidation of pressed particle aggregates known generally as 'sintering'.

The simplest sintering mechanism is that in a monophase ceramic with transport of matter by solid-state diffusion from particle-contacts along grain-boundaries to inter-particle pores. The thermodynamic 'driving-force' for transport is the reduction in free surface energy associated with porosity. To accelerate the sintering process or to achieve a theoretical density (i.e. zero porosity) ceramic it is common practice to make impurity additions to the particle compact. The role of these sintering catalysts is obscure in many systems; various hypotheses involving enhanced diffusion, modified interfacial/surface energy ratios or grain-boundary 'pinning' have been proposed. However, the better spatial resolution of microscopic and analytical techniques over the past decade has established that monophase solid-state sintering is relatively rare. Frequently, sintering additives are the basis for minor intergranular liquids which provide rapid transport paths during densification by a solution-reprecipitation mechanism.

Liquid-phase sintering (l.p.s.) assumes a greater importance in systems with low diffusivity or those susceptible to decomposition. These are normally ceramics in which the major phase has a high degree of covalent bonding such as silicon nitride or silicon carbide, which are prominent examples of the new engineering ceramics. A number of oxides also utilise l.p.s. in fabrication; the restriction in sintering temperature required to develop appropriate phase-mixtures of ZrO_2 polymorphs in some of the new 'toughened' ceramics dictates the use of silicate impurity-based liquids.

The consequence of liquid phase sintering is the presence of intergranular residues which inevitably impair mechanical properties. This processing constraint is a central problem in the ability to microstructurally engineer ceramics for elevated temperature use. Many sintering liquids are silicate based and are

transformed to the glassy state on cooling. Procedures for minimising the intergranular glass volume or tailoring glass compositions for subsequent crystallisation are described below in relation to the new high-temperature nitride and oxynitride ceramics. Heat treatment procedures are also used to induce solid-state phase transformations in various 'toughened' ZrO_2-based ceramics. The capacity for significant property modification via. post-sintering phase-transformations has, in comparison with metallic materials, only recently been identified and is typified by some of the new ceramics.

3. THE REQUIREMENT FOR NEW ENGINEERING CERAMICS

The most rewarding but difficult areas for application of the new structural ceramics are within energy-conversion systems such as the advanced gas-turbine or the turbo-compound diesel engine. The main attraction of ceramics in heat-engine design is the ability to operate at higher temperatures with the elimination or reduction of forced cooling. This translates into greater thermodynamic efficiency and hence decreased fuel consumption (Table 1). Supplementary advantages over cooled metallic alloy components are the reduced inertia (due to a density reduction of 0.5x), important in higher-performance engines. Ceramics, unlike high-performance alloys, rarely contain 'strategic' elements and frequently are composed of the light elements which are abundant in the earth's crust and atmosphere.

The more mundane applications areas for ceramics rely not on their ultimate temperature capability but on their hardness, wear resistance and chemical durability. These are also the main areas of current application whilst the more difficult heat-engine applications remain at a development stage.

Some ceramic component applications are illustrated in Fig.2. Fig.2a illustrates the best-known use of nitride ceramics in high-speed metal cutting under conditions where tool-tip temperatures subject the conventional tungsten-carbide inserts to rapid wear. A comparison of tool lifetimes in the cutting of cast iron is made in Fig.2a.[3] The benefits in machining of certain superalloys are even greater.[4] In this application there is evidence of tip temperatures in excess of 1000°C such that the retention of hardness and wear resistance at elevated temperatures is also important. Similar properties are required in applications involving metal-forming (e.g. extrusion dies) or as high-temperature bearings and seals (Figs.2b). Thermal shock resistance and chemical durability are important in molten metal handling.

The new ceramic applications, especially those in heat-engines have been made possible by the development of ceramics with an appropriate combination of properties. Two

Fig. 1. Scanning electron micrograph of the thermal-protection layer of a space shuttle (courtesy Dr. D J Green).

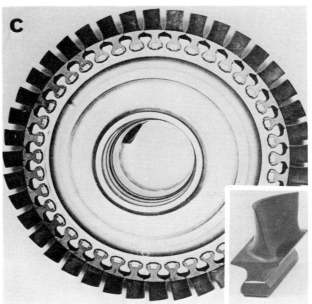

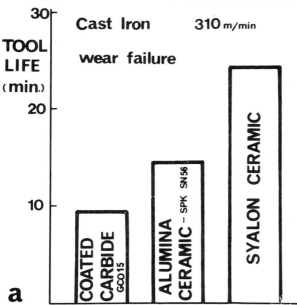

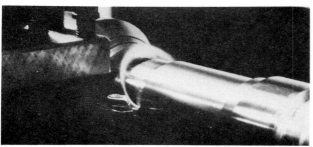

a

Fig. 2. Ceramic components, illustrating some of the existing and potential applications for new engineering ceramics :
(a) Metal machining, with an example of improved tool life for a Syalon ceramic tool-tip.
(b) Various seals, bearings, crucibles and tube-drawing mandrels in Syalon ceramic (courtesy Lucas-Cookson-Syalon).
(c) (d) Turbine blades and shroud rings in Syalon ceramic (courtesy Rolls Royce and Lucas-Cookson-Syalon).

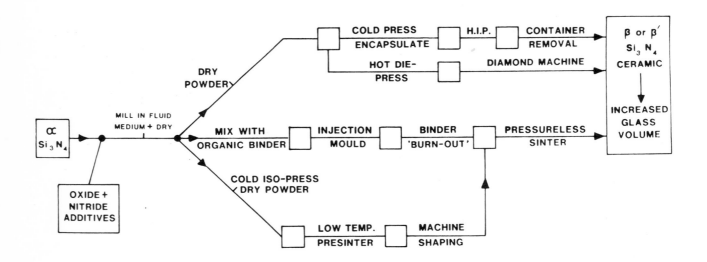

Fig. 2 (e) Reciprocating engine components
(valve components, diesel inlet port,
etc. – courtesy Lucas–Cookson–Syalon).
(f) Turbo–charger rotor.

Fig. 3. Fabrication routes for Si_3N_4 ceramics.

main types of monolithic ceramic may be identified:

(i) Ceramics containing Si_3N_4 or SiC as major phases.

(ii) Ceramics in which ZrO_2 is a major or dispersed phase.

Type (i) ceramics are the main contenders for high-temperature applications (>1000°C) in heat engines and have a property combination of high modulus, hardness, creep resistance, low thermal expansion and hence thermal shock resistance. Good oxidation resistance is conferred via a 'self-healing' SiO_2 based oxidation layer. These properties are intrinsic to the major phases βSi_3N_4 or the substituted 'Sialon' derivatives ($\beta'Si_{3-x}Al_xO_xN_{4-x}$) and either the cubic β SiC or hexagonal αSiC. However, significant property variations are introduced via minor intergranular phases and other microstructural differences which are a consequence of the fabrication process. These microstructure/property relations are discussed in Section 4. Examples of heat-engine components made from this type of ceramic are shown in Fig.2e-h.

Monolithic ceramics suffer from the problem of low fracture-toughness and hence are sensitive to the presence of stress concentrations from point loading and from surface or internal flaws of a size which would be unimportant for metallic alloys. Hence one of the most important ceramic developments in the past decade has been the evolution of a range of ceramics of type (ii) - based on ZrO_2 toughening. These are focussed on the initial discovery[5] of a large increment in strength for the cubic-stabilised ZrO_2 polymorph when heat-treated to contain precipitates of the lower-temperature tetragonal form. This phenomenon has been ascribed to transformation of the metastable tetragonal precipitates to the monoclinic form in the stress field of a propagating crack. The transformation is martensitic and the shapechange is a source of internal stress which modifies the stress field of a crack such that a higher applied stress is required for fracture.[6] It is now recognised that the toughening may be a superposition of different mechanisms, such as crack deflection or microcracking associated with precipitates in addition to stress-induced transformation. The dominance of particular mechanisms depends on the microstructure; in addition to the tetragonal precipitate-containing form (called partially stabilised zirconia - PSZ) high strength and toughness may be achieved with a fine-grained, fully tetragonal, polycrystalline structure (TZP). ZrO_2-toughening may also be applied to other ceramic matrices, such as Al_2O_3, which contain a dispersion of tetragonal or monoclinic zirconia particles.

Apart from enhanced toughness, ZrO_2-based ceramics have emerging applications where their low thermal-conductivity and compatibility of thermal-expansion with metallic alloys is especially valuable. For example, in development of heavy-duty uncooled (adiabatic) diesel engines when recovery of heat-energy in exhaust gases is required (e.g. using the turbo-compound principle) the thermally-insulating properties of ZrO_2 ceramic coatings on cylinder walls, exhaust manifolds etc. have been used. ZrO_2 coatings are also in use in turbine combustors but with the proposed increases in inlet temperature ZrO_2-based ceramics, either as coatings or in the 'toughened' monolithic form, will find little application due to their instability on high-temperature cycling.

Finally, the emergence of a new class of ceramic in the past few years has aroused considerable interest due to enhanced fracture-toughness. This is the ceramic/ceramic composite which is normally that of oriented fibres or whiskers within a ceramic matrix. The type of composite most intensively studied is based on a silicate glass or glass-ceramic matrix[7] and hence has limited high-temperature capability. High-quality fibres are not easily available; those most frequently used are SiC or carbon. The development of low-temperature infiltration techniques for refractory matrices, hence avoiding fibre/matrix reactivity during sintering, appears to have promise[8] but the long-term high-temperature properties of these composites have yet to be demonstrated.

4. THE Si-BASED CERAMICS
4.1. Si_3N_4; fabrication and microstructure

A range of Si_3N_4-based ceramics is exemplified in Table II, identified by acronyms frequently used in the literature. Apart from the low- density reaction-bonded form (RBSN), they all contain βSi_3N_4 as a major phase (>90%). 'Syalon' ceramics contain the derivative β' phase but normally at a very low substitution level such that its intrinsic properties are unchanged. Hence commercial pressureless-sintered products, such as the Syalon ceramics are included within the sintered silicon nitride (SSN) category. They are distinguished by the use of lower-melting ternary eutectic sintering liquids of the type $Y_2O_3-Al_2O_3-SiO_2$, which result in higher sintered densities than the binary type ($Y_2O_3-SiO_2$) or the latter require hot die-pressing (HPSN) or hot isostatic pressing (HIPSN). In all processes the SiO_2 component of the sintering liquid derives from that present on the surface of $\alpha-Si_3N_4$ - the main ingredient of the initial fine-particle mixture. The use of $\alpha-Si_3N_4$ is beneficial not only for its fine particle size and microhomogeneity in SiO_2-distribution but also in providing a chemical component of 'driving-force' for the solution-reprecipitation sintering reaction.

Alternative ceramic-fabrication sequences in achieving shaped components are exemplified in Fig.3. In general the application of pressure during sintering requires smaller liquid volumes and hence reduced silicate

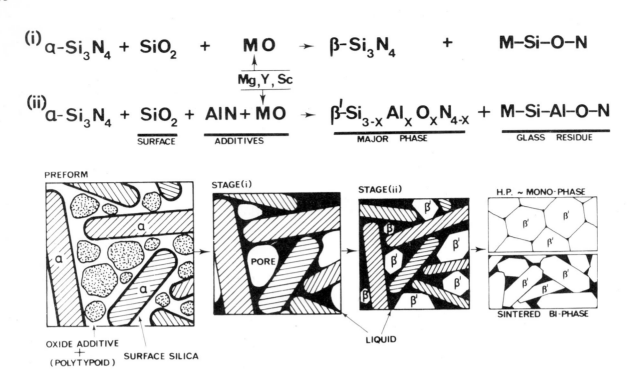

(i) $\alpha\text{-Si}_3\text{N}_4 + \text{SiO}_2 + \text{MO} \rightarrow \beta\text{-Si}_3\text{N}_4 + \text{M–Si–O–N}$

Mg, Y, Sc

(ii) $\alpha\text{-Si}_3\text{N}_4 + \underbrace{\text{SiO}_2}_{\text{SURFACE}} + \underbrace{\text{AlN} + \text{MO}}_{\text{ADDITIVES}} \rightarrow \underbrace{\beta'\text{-Si}_{3-x}\text{Al}_x\text{O}_x\text{N}_{4-x}}_{\text{MAJOR PHASE}} + \underbrace{\text{M–Si–Al–O–N}}_{\text{GLASS RESIDUE}}$

Fig. 4.   Sintering reactions and the evolution
of microstructure for HPSN and SSN
ceramics.

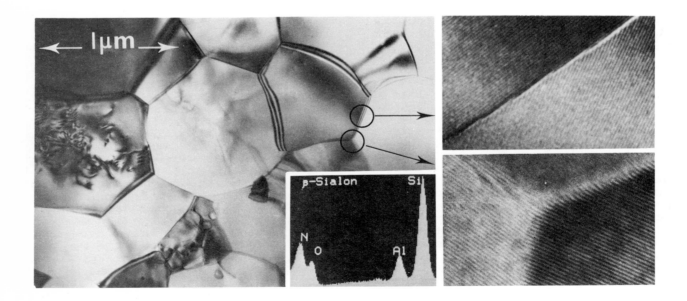

Fig. 5.   The microstructure of a hot-pressed
Si$_3$N$_4$ ceramic, with the absence of
residual intergranular glass revealed
by high-resolution 'lattice' imaging
in TEM.

8

glassy residues. However, the formation of high-density complex shapes by HIP necessitates encapsulation (e.g. with a high-SiO_2 glass) to isolate the ceramic from the high-pressure gas (200MPa) and enable pore closure. HIP also requires a preformed shape via cold die-pressing, injection-moulding, slip-casting or 'green'-machining before encapsulation. These shaping processes are also necessary for pressureless- sintering, normally carried out in 1 atmosphere of N_2 gas at 1700-1800°C. In all cases the sintering shrinkage of 20-30% presents a major problem in dimensional tolerance of finished components.

Sintering reactions and evolution of microstructure are shown schematically in Fig.4 for a hot-pressed (HP) product with minimal liquid and a pressureless-sintered (β-liquid) bi-phase microstructure. Electron-micrographs of these extremes of micro- structure are exemplified by 'Sialon' ceramics (Figs. 5 & 6). The H.P. sialon is a monophase ceramic sintered with a small transient liquid volume (in this case based on $MgO-Al_2O_3-SiO_2$). The P.S. ceramic has 5-10% of a $Y_2O_3-Al_2O_3-SiO_2$ eutectic liquid which cools to the glassy state as a semi-continuous matrix for β'Si_3N_4 crystals. However, in ceramics designed for high-temperature application (>1000°C) it is necessary to increase the glass-softening point and, preferably, to induce total crystallisation of the glass matrix. In β' ceramics control of glass matrix and viscosity is achieved by substitution of part of the added Al_2O_3 by AlN (added as a 'polytypoid' phase containing substituted Si and O). The increased N/O ratio in the glass also makes it susceptible to crystallisation of the mixed oxide 'yttro-garnet' (YAG) rather than a silicate phase. Excess Si and N is accommodated in β' and, with carefully tailored initial compositions and post-sintering heat-treatments, the product is a fully-crystalline bi-phase (β'+YAG) ceramic typified by the '201' type of Lucas-Cookson-Syalon.

The earlier commercial silicon nitride ceramics were based on simultaneous nitriding and sintering of compacted silicon powder (reaction-sintering). These RBSN's (Table II) have high porosity levels and hence low strength, but exhibit negligible shrinkage during reaction sintering at 1400°C. More recently a hybrid fabrication route of reaction-sintering followed by high-temperature sintering has been explored (SRBSN - Table II). Final microstructures are generally similar to SSN but the smaller total shrinkage (10%) is beneficial in ceramic shaping.

4.2. Si_3N_4; properties

A survey of mechanical behaviour for selected fully dense high-strength nitride ceramics is contained in Figs.7-10. There is a division in mechanical behaviour from that of a brittle, high modulus, solid at temperatures below 1000°C to a solid which undergoes creep deformation and time-dependent fracture at higher temperatures. The transition is linked to the softening temperature of intergranular glassy residues which increases with N/O ratio and with metallic additive valence (e.g. MgO to Y_2O_3).

In the low-temperature range brittle failure is initiated at pre-existing stress-concentrating flaws. The bend strength (modulus of rupture - M.O.R.) is typically in the range 600-1000 MPa and, as in all brittle solids, is determined by the combination of an intrinsic microstructure-dependent parameter K_{1c} (the fracture-toughness) and the maximum flaw 'size' ∿c via the relation $MOR = \frac{1}{Y} \frac{K_{1c}}{\sqrt{c}}$

is a geometrical 'flaw shape' parameter. This ideal relation is plotted in Fig.7 and gives an approximate range of flaw sizes in the more recent high-strength nitrides. Values of K_{1c} used in Fig.7 are typical of the Syalon range of pressureless-sintered ceramics in which β' grains, crystallised in a liquid matrix, have an elongated prismatic shape.[9] HP ceramics generally have lower values (K_{1c} 3-5 MPa m$^{\frac{1}{2}}$) due to their more equiaxed grain shapes, thus giving a smaller intergranular crack-deflection/'pull-out' component of fracture-toughness. In monolithic nitride ceramics there is little prospect of large increases in K_{1c} such that enhanced M.O.R., together with reduced statistical spread (measured via the Weibull modulus)[10] must come from improved purity, homogeneity in powder processing and sintering to theoretical density. This is relatively difficult in multi-component nitride/oxide systems. Impurity inclusions normally derive from fragments of ball-milling media (e.g. Al_2O_3) or from transition elements (e.g. Fe) within Si_3N_4 powders which form silicides on sintering.

The achievement of MOR values above 1 GPa requires a maximum flaw size <30μm according to the idealised relation in Fig.7. However, it is now realised that surface machining produces an array of surface flaws which contain residual stresses due to a mismatch between the plastically-deformed surface layer and the underlying ceramic.[11] The residual stress supplements the applied stress in crack-opening such that flaws which are much smaller than the stress-free critical size, defined by Fig.7, may result in failure. The failure stress is then a function of the residual stress, determined by severity of machining and ceramic hardness, in addition to K_{1c}.[11]

The transition to 'high'-temperature mechanical behaviour in nitride ceramics containing intergranular glass residues is typified by a precipitous fall in MOR, normally accompanied by a transient rise in K_{1c} (Fig.8). The transition is linked to the glass softening temperature with grain-boundary sliding and viscous deformation of glass bridging crack surfaces all contributing to enhanced K_{1c}. In this regime fracture is not always initiated at pre-existing flaws but, at lower stresses, is associated with the accumulation of creep-cavitation damage i.e. failure is time and strain-dependent.

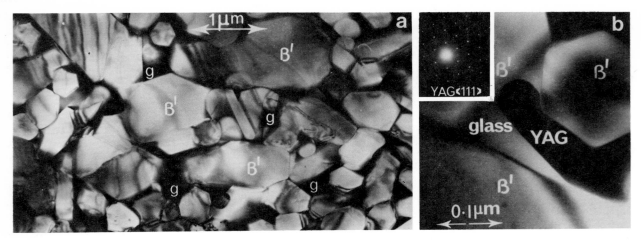

Fig. 6. The microstructure of a SSN ceramic, composed of faceted Si_3N_4 crystals within a ternary- eutectic $(Y_2O_3-Al_2O_3-SiO_2)$ glass matrix containing nitrogen. At high-nitrogen content the matrix crystallises to yttro-garnet (YAG) - Fig.6b.

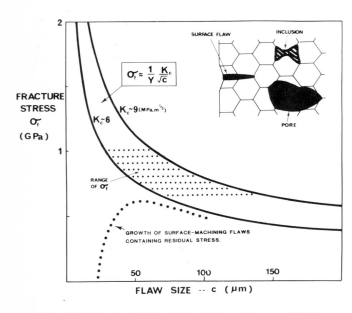

Fig. 7. A simplified fracture-stress (MOR) - flaw size relation plotted for a range of fracture-toughness (K_C) measured in the best SSN ceramics and compared with the experimentally-measured range of MOR. Machining flaws contain residual stresses which may result in a precursor crack growth of small flaws to a reduced failure stress (dotted curve).

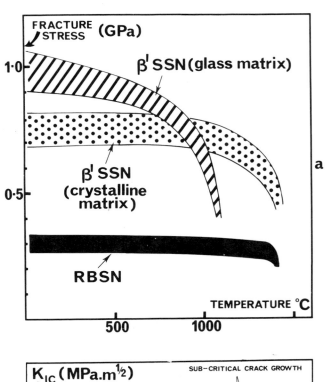

Fig. 8. MOR and K_{1C} - temperature relations for Si_3N_4-based ceramics. The susceptibility to sub-critical crack growth is dependent on elimination of residual glass junctions which form cavity nucleation sites.

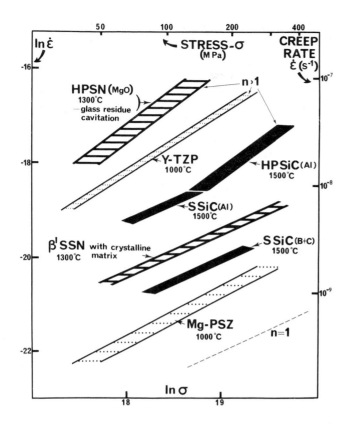

Fig. 9. A comparison of pseudo-steady-state creep rates for a range of Si-based and ZrO$_2$-toughened ceramics within differing temperature intervals. The stress-exponent (n) of the creep rate is frequently=1, indicative of a diffusion-controlled mechanism.

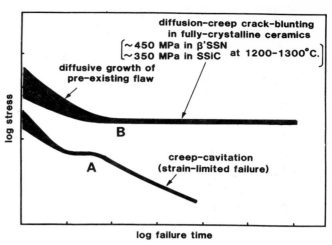

Fig.10. Stress(creep)-rupture behaviour for Si-based ceramics. A high-stress crack-blunting theshold is typical of completely crystalline ceramics undergoing diffusional creep (type B) and is predicted from crack-growth data (Refs.13,14). Type A behaviour is typical of ceramics containing glass residues which undergo creep-cavitation; a limited blunting threshold is followed by fracture due to coalescence of creep-damage (strain-limited failure).

There have been significant advances in the control of high-temperature deformation and fracture via the increased flexibility of micro-structural engineering in the Si-Al-O-N system. This is based on the suppression of creep cavity nucleation when grain-junction glass is below a critical size and is achieved by adjustment of composition and post-sintering heat-treatment to approach the fully crystalline state discussed in 4.1. This microstructure change is also associated with a reduced creep rate with a 'Coble' mechanism controlled by grain-boundary diffusion (Fig.9). Terminal steady-state creep rates at 1300°C are similar for H.P. monophase and P.S. bi-phase ceramics.[12] Creep-rupture data, available only for a limited range of nitride ceramics, indicate a behaviour summarised in Fig.10. At high stresses failure originates from pre-existing flaws with crack-propagation in fully-crystalline ceramics which has been modelled via diffusional growth.[13,14] There is

increasing evidence for a threshold stress, below which the 'blunting' of pre- existing flaws occurs by diffusional creep or viscous flow of intergranular glass, depending on the microstructure. This blunting threshold has been identified in fully crystalline Si-Al-O-N ceramics from crack-velocity/K$_1$ data[13,15] and from creep-rupture tests the plateau is believed to extend to very long times in the absence of creep-cavitation. In those ceramics (typified by commercial HP nitrides of the type NC132) where intergranular cavities nucleate below the threshold stress failure is strain controlled via coalescence of cavitation-damage. This behaviour has also been identified in oxides[16] and may be of general application to ceramics. The transition from type A to type B creep-rupture behaviour is critically dependent on micro-structure (i.e. intergranular glass volume and composition) but may also be a function of temperature.

High temperature oxidation-resistance in

Si_3N_4 ceramics is conferred by a passive SiO_2-rich oxidation film, normally in the viscous liquid state at high temperatures. The intrinsic oxidation rate of β Si_3N_4, controlled either by oxygen diffusion through SiO_2 or by the solution rate of Si_3N_4 in SiO_2, is very low even at temperatures >1400°C. The main problem is due to out-diffusion of metallic ions from the intergranular phase which reduces the SiO_2 film viscosity and hence increases the oxygen diffusion rate or Si_3N_4 solubility.[17] For PS ceramics in which the glassy-matrix forms a large reservoir and rapid diffusion path for metallic ions, oxidation resistance above 1200°C is poor. In addition the metallic-ion rich films are susceptible to crystallisation and spalling, especially on thermal cycling. Matrix-phase crystallisation is a partial solution to this problem; for example, YAG or $Y_2Si_2O_7$ matrices in Syalon ceramics confer 1300°C oxidation rates comparable with HP monophase ceramics (Fig.11). Above 1300°C oxidation rate is sensitive to matrix crystal chemistry; most metallic oxides, such as YAG react with SiO_2 to form a eutectic liquid which is accompanied by reversion to high oxidation rates.[18] An extension of application temperature to 1400°C is possible by surface coating (e.g. with CVD Si_3N_4 or SiC) or via controlled sub-surface transformation of the type :

$$Si_3N_4 + SiO_2 \rightarrow 2Si_2N_2O$$

following out-diffusion of metallic ions from the intergranular glass.[18] Si_3N_4/Si_2N_2O phase combinations (or their substituted derivatives β'+O') have excellent oxidation resistance. Fig.11.

4.3. SiC ceramics

Ceramics in which the hexagonal αSiC or cubic βSiC is a major phase have increasing potential as engine components in competition with Si_3N_4. Important steps in the development of dense SiC ceramics were, firstly, the identity of sintering additives for hot-pressing [19] and, later, the ability to pressureless sinter ultrafine βSiC powders.[20] The broad division of ceramic type, presented in Table III, is similar to that for Si_3N_4-based ceramics, viz hot-pressed (HPSiC), pressureless-sintered (SSiC) and a type of 'reaction-sintering' between a graphite-containing preform and silicon introduced as a liquid or vapour (siliconized SiC).

Sintering additives for HPSiC are either B+C (added as B_4C) or Al_2O_3. The latter results in microstructures and properties similar to HPSN, dominated by intergranular (alumino silicate) residues at high temperatures. Fully dense fine (1-2μm) grain-size microstructures provide high MOR values (600-800 MPa) below 1000°C with subsequent rapid degradation beyond the (1200°C) silicate softening temperature (Fig.12). The function of B+C additives (0.5-1%) in both HPSiC and SSiC ceramics is believed to be the removal (via C) of the SiO_2 surface contaminant on SiC powders and the catalysis (via B) of

surface/grain-boundary diffusion, hence accelerating densification. Pressureless-sintering requires slightly higher temperatures coupled with ultra fine, low oxygen, SiC powders (both α and β powders and products have been used - Table III). Grain-sizes and porosity levels are higher than for Al_2O_3 additives due to the absence of more efficient intergranular sintering liquids which also inhibit grain growth. As a consequence low-temperature strength is inferior but strength is retained to higher temperatures than for Al_2O_3-additive ceramics or HPSN and SSN (Fig.12).[21] The poor low-temperature strength of SSiC is a combination of large flaw size and relatively low fracture-toughness. Typical values for K_{1c} are 3-4 MPa $m^{\frac{1}{2}}$, probably due to grain size and lack of anisotropy in grain shape in comparison with SSN.

The superior high-temperature properties of (B+C additive) SSiC reflect the absence of liquid-sintering residues. B+C may be present only as undetectable intergranular segregated atoms following an essentially solid-state sintering process. Equivalent creep rates to SSN are attained at much higher temperatures (Fig.9) and the linear stress dependence indicates a diffusion-controlled mechanism. There are various interpretations of the diffusion process (lattice or grain boundary) and rate-controlling species which appear to be a function of the sintering additives.[22,23] The appearance of a non-linear (frequently non-integral) stress dependence and visible creep-cavitation in Al_2O_3-additive ceramics is explicable via their inter-granular residues and parallels that of inferior HPSN and SSN ceramics, although normally at higher temperatures due to the higher alumino-silicate glass viscosities.[24]

SiC ceramics are less susceptible to time-dependent failure (creep-rupture) than most Si_3N_4-based ceramics up to 1600°C. This stems from the non-cavitating creep-behaviour of (B+C) SSiC such that the low-stress 'strain-limited' failures (Fig.10) are absent. Like the fully-crystalline Si_3N_4 ceramics there is a narrow range of stress within which time-dependent failure is measureable above a crack-blunting threshold (Fig.10). Stress intensity exponents(n) in the crack-velocity relation $V=AK^n$ normally exceed 30 for tests in vacuum or argon[25,26] characteristic of a nearly time-invariant fracture-stress. In oxidising atmospheres the exponent increases, suggesting a crack-growth mechanism associated with an intergranular oxidation-reaction, i.e. stress-corrosion,[27] which is probably sensitive to the segregated species at grain-boundaries. Creep-rupture data for SSiC ceramics is not directly comparable with that for SSN ceramics, due to their substantially lower instantaneous fracture stress. However, in the 1300-1500°C interval current SiC ceramics have a superior performance.

Siliconized SiC ceramics have, until recently, been associated with lower values of MOR and a rapid loss of strength at 1400°C, associated with the melting of unreacted Si

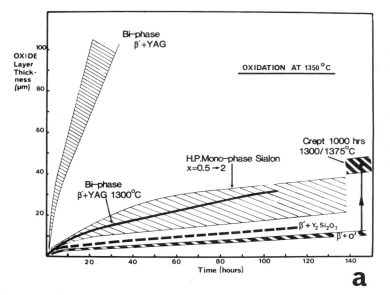

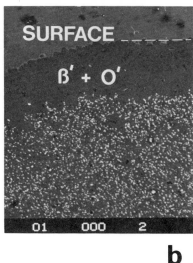

Fig.11.   A comparison of oxidation kinetics for
β′ HPSN and β′ SSN ceramics with
different phase combinations within
near-surface microstructures.   The
β′+O′ (Si$_2$N$_2$O) data corresponds to a
surface transformation layer
illustrated in 11b.

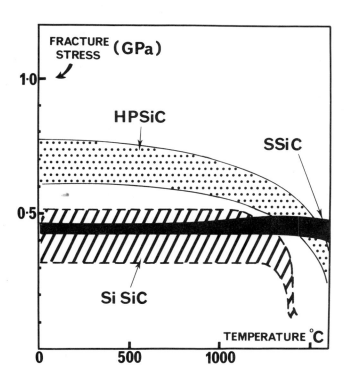

Fig.12.   Bend strength (MOR) - temperature
relationships for various SiC-based
ceramics (see table III).

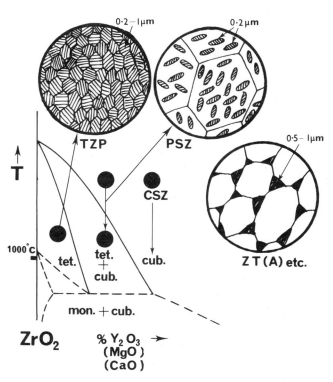

Fig.13.   Phase equilibria and related
microstructures for ZrO$_2$-toughened
ceramics.

(Fig.12). A new processing technique, in which a porous preform is made by carbonization of a liquid polymer prior to Si-infiltration, has achieved MOR values comparable with the best SSN ceramics.[28] Shaping of components may be carried out at the polymer-precursor stage due to predictable shrinkage on carbonization. Excess Si still provides the limit to high-temperature application, but creep-rupture properties are excellent to ~1300°C.

Apart from their inferior MOR and fracture toughness, SiC ceramics suffer from a relatively poor thermal shock resistance compared to Si_3N_4 ceramics. This is due to a higher thermal expansion coefficient (α) and elastic modulus(E), interpreted via a thermal stress resistance parameter such as $R = \dfrac{\sigma(1-\nu)}{\alpha E}$, ($\sigma$ = M.O.R., ν = Poisson's ratio). This is one of the main reasons why Si_3N_4-based ceramics have been favoured in turbine applications for highly-stressed components. However, SiC ceramics have found application in combustors, in which temperature variations are not severe, utilising their superior thermal-diffusivity in eliminating 'hot-spots'. The absence of intergranular glassy residues in SSiC results in a superior oxidation resistance compared to SSN ceramics, especially above 1350°C (Section 4.2).

5. ZrO_2 TOUGHENED CERAMICS
5.1. Constitution and Microstructure

The constitution of PSZ and TZP ceramics in relation to phase equilibria is sketched in Fig.13. The cubic polymorph of ZrO_2 is 'stabilized' by the extension of its transformation temperature (cubic - tetragonal) to lower temperatures when in the form of a solid-solution with various metal oxides. MgO CaO and Y_2O_3, which are the most frequently used stabilizing elements, produce different phase-diagrams but share a common feature in the separation of C and T polymorphs by a two-phase (C+T) field which is the basis of PSZ ceramics.[29]

A typical PSZ, alloyed with 8-10 mol.% MgO, CaO or Y_2O_3 would be sintered in the 1650-1850° range to produce large (~100μm) grains of cubic solid solution, cooled and heat-treated in the 1100-1450°C interval to produce coherent precipitates of the tetragonal phase (Fig.14). Precipitate morphology depends on stabilizing oxide (lenticular for MgO, cuboid for CaO, platelet for Y_2O_3) and heat-treatments are critical in relation to precipitate particle size (normally ~0.2μm) such that the constrained tetragonal particles will transform to the monoclinic stable form, at lower temperatures, only in the presence of the crack stress-field.[30] 'Overaging' causes a spontaneous transformation of the larger particles on cooling whilst 'underaging' inhibits crack-tip transformation.

TZP ceramics have lower solute oxide additions such that they lie in the tetragonal single-phase field when sintered in the 1350-1500°C interval (Fig.14c and Table IV). Y_2O_3 (2-4 mol.%) is the favoured solute oxide

and there is a critical grain-size for stabilization of a fully tetragonal structure to lower temperatures. This is normally in the 0.2-1μm range but depends on Y_2O_3 concentration and residual grain-boundary phases.[31] These residues are derived from impurity silicates which, together with Al_2O_3 additions and the Y_2O_3 solute, are the basis of a silicate sintering liquid analogous to that used for Si_3N_4 ceramics. Liquid phase sintering is a necessity in the lower-temperature tetragonal phase field.

The dispersion of tetragonal-stabilized ZrO_2 within other ceramic matrices is the basis of a third class of microstructure which exhibits ZrO_2-toughening phenomena (Fig.13). The tetragonal particles may be either intra or intergranular dispersions but the latter, exemplified by zirconia-toughened Al_2O_3 (ZTA - Fig.14d) is most effective and conveniently derived from normal ceramic processing.[32] The dispersed tet-ZrO_2 particle sizes are also critical in relation to thermal stability and crack-induced transformation. They are in the 0.2-1μm range, the particular optimum value being a function of matrix property and impurity level; for example, in ZTA the critical size is 0.5μm with a related (0.5-5μm) Al_2O_3 grain size.[33]

5.2. Properties and Limitations

Some of the low-temperature properties which emphasise the potential of ZrO_2-toughened ceramics are listed in Table IV. Mean values of MOR and fracture-toughness are far superior to a cubic stabilized zirconia (CSZ) and there is an important increment above the high-temperature monolithic Si_3N_4 and SiC ceramics. The high values of K_{1C} stem from both transformation of metastable tet-ZrO_2 in the crack stress-field and from transformation-induced microcracking, the relative importance of these mechanisms varies with type of microstructure (PSZ, TZP or ZTA). MOR is limited by K_{1C}, but also by flaw population via a 'Griffith-type' relation (Fig.7). Hence the highest strengths are recorded for TZP (a remarkable 2500 MPa) due to a combination of K_{1C} and controlled processing of fine powders, avoiding both porosity and heterogeneity.

The impressive low-temperature properties are not retained to elevated temperatures for two main reasons. Firstly, the thermodynamic driving force underlying the transformation-toughening mechanism decays as the equilibrium transformation temperature is approached (~1000°C). This is shown in Fig.15 with an example of a predicted and measured strength-temperature relation for a ZTA ceramic.[32] Secondly, the need for liquid-assisted sintering, especially for TZP, severely impairs high-temperature strength and creep resistance due to normally-glassy intergranular residues (Fig.9). An additional problem concerns thermal cycling; although transformation-toughening is beneficial in reducing thermal-shock damage, there are a number of microstructures (especially Mg-PSZ)

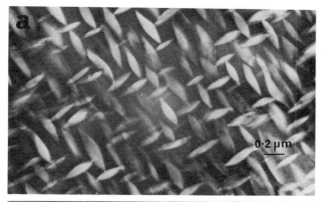

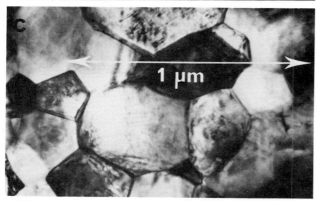

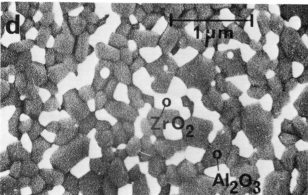

Fig.14. Electron micrographs of ZrO$_2$-toughened ceramics :
(a) Mg-PSZ with transformation of initially-tetragonal precipitates, adjacent to a crack, revealed by dark-field imaging in (b) using monoclinic reflections. (Courtesy Professor A H Heuer; (b) from Schoenlein and Heuer in 'Fracture Mechanics of Ceramics <u>6</u>, Plenum Press 1983, (a) from D L Porter, PhD thesis, Case Western Reserve University).
(c) Y-stabilized TZP.
(d) ZT-Al$_2$O$_3$ containing intergranular tetragonal ZrO$_2$ (back-scattered SEM - courtesy Dr. D J Green).

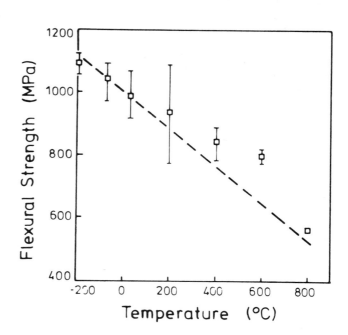

Fig.15. MOR-temperature relation for a ZT-Al$_2$O$_3$, illustrating the reduction in strength related to an approach to the equilibrium (tetragonal-monoclinic) transformation temperature (after Ref.6 and 35).

which suffer thermally-induced irrecoverable transformation from tetragonal to monoclinic ZrO_2 when cycled above 800°C. However Y-TZP is stable up to 900°C and hence is favoured in thermal-barrier applications utilising the low thermal conductivity of ZrO_2 and its metallic-range thermal expansion.

There have been various alloy-design strategies used in attempting to extend the application temperature for ZrO_2-toughened ceramics. Using HfO_2 as a stabilising additive equilibrium transformation temperatures up to 1400°C have been achieved, but there are problems in control of sintered microstructure to retain a critical size for the tetragonal phase (e.g. in Hf-ZTA).[34] Alternative strengthening mechanisms, for example, precipitation within the matrix of ceramics such as ZTA have been explored[32] together with a refinement or removal of grain-boundary residues. The latter procedure normally impairs sinterability, especially in TZP but also in PSZ ceramics where impurity SiO_2 (~0.05% on commercial ZrO_2 powders) is recognized as an important liquid sintering catalyst. The most promising high-temperature strengthening mechanism is that using whiskers (normally SiC) dispersed within matrices such as Y-TZP, ZT-Al_2O_3, ZT-mullite or ZT-cordierite.[35] These are essentially a class of ceramic-matrix composites discussed in the next section.

6. CERAMIC COMPOSITES
6.1. Strengthening principles

The principle of load-transfer to high-modulus, high-strength fibres within polymer or metallic matrices is well-established in achieving high-strength, fracture-tough materials. Carbon fibre reinforcement of glass was demonstrated more than 10 years ago[36] but further developments using ceramic matrices have been inhibited by fibre damage or reactivity during processing and a severe limitation to high-temperature properties from atmospheric oxidation of carbon fibres.

A resurgence of interest in composites with ceramic matrices is due partly to the problems of engineering design with brittle monolithic ceramics but also to the availability of a variety of reinforcing phases, for example SiC, Si_3N_4 and Al_2O_3 single crystal whiskers or SiC produced from a polymer precursor as a continuous fibre. A particular stimulus has been the work of Prewo et al at United Technologies Research Center, U.S.A., in demonstrating high-strength and fracture-toughness for silicate glass ceramics reinforced with high-strength (>2000MPa) continuous SiC fibres[7] (a Japanese 'Nicalon' source, which is a microcrystalline SiC interspersed with SiO_2 and free carbon).[38] The approach to an optimum mechanical response of Nicalon-reinforced lithium-aluminium silicate (LAS) glass-ceramic has confirmed the strengthening principles for semi-continuous fibres, illustrated in Fig.16. The important mechanism is that of load-transfer following matrix-cracking which results in a deviation from composite elastic behaviour (Stage I) to a reduced modulus associated with elastic extension of fibres (Stage II). A maximum stress (the UTS) occurs with fibre-bundle failure which may be followed by fibre 'pull-out' (Stage III) which contributes to the total work-of-fracture. The important characteristics of this ideal response are :

(i) A high failure strain and tolerance of overstressing (such as thermal shock or impact loading) which would cause instantaneous fracture in a monolithic ceramic.

(ii) An insensitivity to matrix flaws and hence component size (unlike the 'weakest-link' failure of monolithic ceramics) due to a limitation to crack-tip stress intensity caused by fibre-bridging of crack surfaces.[16]

This ideal composite behaviour is only achieved for unidirectional semi-continuous fibres which are weakly bonded to the matrix. Fibre-matrix reaction which induces primary interface atomic bonding results in a monolithic-type of response. Formation of weak interfaces is a function of processing (e.g. temperature at which initial matrix glass infiltration occurs) and chemical compatibility. There is evidence of a 50nm amorphous carbon-rich interfacial layer in certain optimal SiC-glass ceramic composites (e.g. SiC-LAS)[39] which provides an appropriate combination of weak interface and interfacial shear resistance during load transfer.

Whisker-reinforced composites are normally fabricated by conventional ceramic powder mixing and sintering techniques but hot-die pressing, used to achieve theoretical density, induces a preferred two-dimensional whisker orientation. The induced property-anisotropy does not extend to the ideal fracture behaviour observed for uniaxial semicontinuous fibre reinforcement. However, there are important increments in strength and fracture-toughness which may be attributed to interfacial micro- cracking and whisker pull-out which is also sensitive to whisker-matrix reactivity.[41] Such mechanisms relate to the enhanced K_{1c} levels ascribed to the anisotropy in βSi_3N_4 crystal growth morphology in SSN ceramics. Single crystal whiskers, such as βSiC, used in composites have large anisotropies, typically 0.2-0.5µm x 5-30µm.

6.2. Phase Combinations and Property Limitations

Semicontinuous fibres of C and βSiC are available as refractory reinforcing phases with high-temperature potential within ceramic matrices. Oxide fibres and whiskers are unsuitable due to reactivity with silicate liquids used either as glass-ceramic matrix precursors or as liquid-sintering aids used during conventional ceramic processing. αSi_3N_4 is unstable in most silicate liquids above 1000°C, being susceptible to the dissolution-reprecipitation reaction utilised in βSi_3N_4 ceramic fabrication (section 4). Since carbon is susceptible to oxidation in the uncoated state within silicate matrices the most recently studied composites are based on βSiC fibres and whiskers.

The problem of infiltration and reactivity during processing has promoted the use of silicate glasses and their crystalline derivatives as suitable matrices for continuous fibres. They may be tailored with a wide range of chemistry, softening point or thermal expansion mismatch with the reinforcing phase. The composites may be fabricated by hot-pressing following infiltration of fibres with a presynthesised glass frit[36,37] or via a sol-gel glass route which has the advantages of improved homogeneity in microstructure and reduced hot-pressing temperature. [42] More refractory matrices have been formed via chemical-vapour-infiltration (CVI), the comparatively low temperature processing then permits a much wider range of both fibre and matrix chemistry.[8,43] The absence of the pre-alignment constraint for whisker-reinforcement permits the use of powder mixing and sintering procedures, for example, βSiC whiskers within Al₂O₃ or Si₃N₄ ceramic matrices.[41,44]

The high values of fracture stress attainable for unidirectional fibre-reinforced SiC-LAS glass-ceramic composites, described above, may be extended to temperatures of ∼1000°C (Fig.18a). However, this high temperature data exposes a number of potential limitations of such composites. Firstly, above the matrix- microcracking stress the fibres are exposed to the environment, which causes premature failure in oxidising atmospheres above ∼700°C (Fig.18b) probably due to reaction with the carbon-rich interfacial layer.[39] Secondly, there is little data yet available on time-dependent deformation, but the transient increase in fracture strength and K_{1c} between 700 and 1000°C is indicative of matrix plasticity due to residual glass. This is normally associated with creep-cavitation failure in monolithic ceramics. The use of more refractory matrices, such as CVI SiC, reveals an ultimate temperature limitation due to irreversible structural changes within the Nicalon SiC fibres.[21] This is consistent with a reduction in isolated- fibre fracture stress by a factor of 2 following heat treatment at 1300°C for 200 hours[42], believed to result from a recrystallisation/ grain growth process within the SiC/SiO₂/C fibre microstructure.

Encouraging data for βSiC whisker-reinforcement of Al₂O₃[41] and Si₃N₄[44] has recently been presented. For hot-pressed Al₂O₃ components with >40 vol.% SiC there is a fracture stress increment over a monolithic Al₂O₃ of 500 MPa which is retained to ∼1200°C (Fig.19). Fracture toughness increases from 4.5 to 8.5 MPa m and there is a remarkable insensitivity to thermal shock up to the 1000°C test limit.[41] Hot-pressed Si₃N₄ composites with 30 vol.% SiC have been fabricated with 600-700 MPa bend strengths, comparable with monolithic ceramics, but combined with higher Weibull modulus and work of fracture.[44] These composites have a supplementary advantage in being electrically conducting and hence may be spark-machined to component shape. As in the case of fibrous-composites, the time or environmental dependence of high-temperature mechanical behaviour has yet to be proven and it is clear that whisker-matrix interfacial reactivity is sensitive to time and temperature of hot-pressing in addition to composition of liquid sintering aid[45] and whisker surface chemistry.[44]

7. FUTURE DIRECTIONS

Each class of structural ceramic surveyed in this paper, although representing a significant advance in the past decade, is observed to have important limitations within differing regimes of temperature. A simplistic survey of this situation is contained in Fig.20 in which fracture stress is plotted as a function of temperature for examples of the monolithic ceramics.

Engineering design with the Si-based ceramics is inhibited by their relatively low K_{1c} and MOR combined with a spread in property (low Weibull modulus) associated with failure from the dominant flaw. Some improvement has been realised for SSN ceramics due to a natural anisotropy in βSi₃N₄ crystal morphology and hence enhanced K_{1c}. Comparatively poor low-temperature properties for SSiC ceramics are compensated by the high-temperature increment over SSN ceramics containing liquid sintering residues. The importance of complete crystallisation of intergranular glass in the latter is well- established in relation to creep rate and stress-rupture. For completely crystalline SSN ceramics the temperature ceiling is currently limited by oxidising atmospheres.

Zirconia ceramics offer a valuable increase in K_{1c} and MOR and the toughening principles are also applicable to a variety of, normally oxide, ceramic matrices. The intrinsic limit to transformation toughening and the presence of intergranular sintering residues inhibits their application at temperatures comparable with the Si-based ceramics.

A quantum jump in the performance of monolithic ceramics would occur in combining the advantages of Si-based and ZrO₂-toughened ceramics at high and low temperatures, respectively. However, the development of an optimal microstructure consisting of tetragonal ZrO₂ intergranular particles of critical size within a SSN or SSiC fine-grain polycrystal with no residual glass phase is difficult to achieve, in view of the differing constraints of sintering temperature and chemistry. Similar difficulties occur in attempting to improve the high-temperature performance of ZrO₂-toughened oxides, such as Al₂O₃, mullite, spinel etc.[35], in which higher-temperature sintering is a consequence of the quest for creep resistance via removal of liquid sintering aids. Such difficulties dictate the alternative strategy involving composite ceramics; for example, SiC whisker dispersions within SSN or oxide matrices to improve low

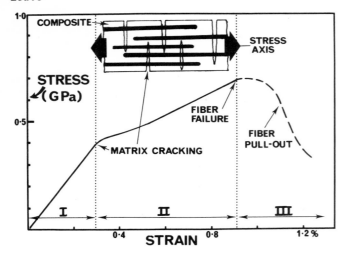

Fig.16. The optimal stress-strain behaviour of a ceramic composite, typified by an LAS glass ceramic containing unidirectional 'Nicalon' SiC fibres (after Ref.7).

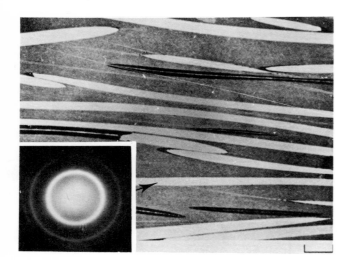

Fig.17. Longitudinal section through a SiC/LAS glass-ceramic (after Ref.7). A transmission electron diffraction pattern (inset) illustrates the microcrystalline state of β SiC which is associated with residual graphite and amorphous SiO_2 in the Nicalon fibers.

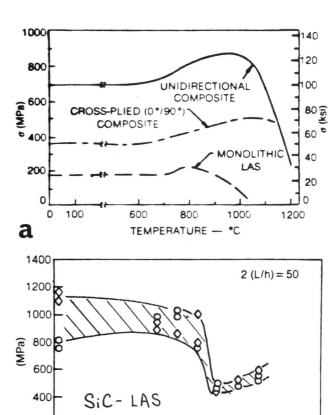

a

b

Fig.18. Fracture stress-temperature relation for SiC/LAS composites :
(a) Inert-atmosphere tests
(b) Unidirectional composite tested in an oxidising environment (Ref.40).

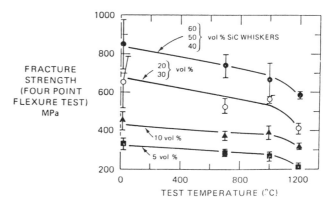

Fig.19. MOR-temperature data for SiC whisker-reinforced, hot-pressed Al_2O_3 composites (Ref.41).

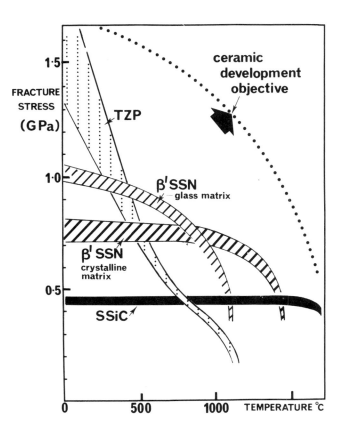

FRACTURE STRESS (GPa)

1·5

1·0

0·5

0   500   1000   TEMPERATURE °C

ceramic development objective

TZP

β'SSN glass matrix

β'SSN crystalline matrix

SSiC

Fig.20. A summary of the limitations of the various types of monolithic ceramic in differing temperature regimes, indicating the need for development of new composite ceramics.

temperature fracture toughness and within ZrO₂-toughened ceramics to improve high-temperature creep (and further enhance low-temperature properties). Both strategies, which are at an early development stage exemplified above[35,41,44], necessitate fabrication via hot-pressing. This is a serious constraint in an already-difficult area of net-shape component fabrication for monolithic ceramics. The fabrication of CVI composites is currently applicable only to simple shapes with semi-continuous fibre reinforcement. Improvement in high-temperature fibre and interface stability, especially in oxidising atmospheres, is an additional requirement.

In conclusion it is important to emphasise the degree of sophistication which has evolved over the past decade in control of ceramic microstructure and properties beyond that of the 'as-sintered' state. In some ZrO₂-based ceramics solid-state precipitation is used in development of microstructures which utilise a martensitic transformation as a toughening mechanism. Improved high-temperature performance in liquid-phase-sintered ceramics is achieved via post-sintering heat-treatments designed to crystallise residual glass matrices and to enhance oxidation-resistance via transformation of surface microstructure. Thus the development of ceramic alloys is paralleled by the earlier development of metallic alloys in which initial chemistry and subsequent thermal treatment are dictated by stress, temperature and environment in service.

ACKNOWLEDGEMENTS:

The author wishes to thank research associates at Warwick and elsewhere who have supplied both published and unpublished Figures for this review. Financial support for research programmes at Warwick which relate to part of this review has been received from Lucas-Cookson-Syalon, Rolls-Royce, S.E.R.C. and the Wolfson-Foundation.

REFERENCES :

1. D J Green, in 'Industrial Materials Science and Engineering; ed. L E Murr, (Marcel Dekker 1983) 89.

2. R N Katz, Mat.Sci. and Engineering 71 (1985) 227.

3. S K Bhattacharyya, A Jawaid and M H Lewis, Proc.XII NAMRI Conf., Michigan Tech.Univ., (1984).

4. S K Bhattacharyya, A Jawaid and J Wallbank, Metalwork.Prod. 126 (1982) 104.

5. R C Garvie, R H Hanninck and R T Pascoe, Nature, 258 (1975) 703.

6. F F Lange, J.Mat.Sci., 17 (1982) 225.

7. K Prewo and J J Brennan, J.Mat.Sci. <u>17</u> (1982) 1201 and 2371.

8. R Naslain and F Langlais, in 'Tailoring Multiphase and Composite Ceramics' ed. Tressler and Bradt (Plenum 1985) in press.

9. M H Lewis and R J Lumby, Powder Metall. <u>26</u> (1983) 73.

10. R W Davidge in 'Micromechanisms of Plasticity and Fracture' ed. Lewis and Taplin (Parsons Press 1983) 167.

11. D B Marshall, A G Evans, B T Khuri Yakub, J W Tien and G S Kino, Proc.Roy.Soc. A385 (1983) 461.

12. M H Lewis, S Mason and A Szweda in 'Proc.Conf. on Non-Oxide Technical and Engineering Ceramics' ed. Hampshire (Elsevier 1985) in press.

13. M H Lewis and B S B Karunaratne in 'Fracture Mechanics for Ceramics, Rocks and Concrete', ed. Freiman and Fuller (ASTM STP745 1981) 13.

14. T J Chuang, J.Amer.Ceram.Soc. <u>65</u> (1982) 93.

15. B S B Karunaratne and M H Lewis, J.Mat.Sci. <u>15</u> (1980) 449 1781.

16. A G Evans, Mat.Sci. and Engineering <u>71</u> (1984) 3.

17. M H Lewis and P Barnard, J.Mat.Sci. <u>15</u> (1980) 443.

18. M H Lewis, B S B Karunaratne, J. Meredith and C Pickering, in 'Creep and Fracture of Engineering Materials and Structures', ed. Wilshire and Owen (Pineridge Press (1981) 365.

19. S Prochazka in 'Silicon Carbide-1973', eds. Marshal, Faust, Ryan, (University of S.Carolina Press 1974) 391.

20. S.Prochazka in 'Ceramics for High-Performance Applications', eds. Burke, Gorum, Katz (Brook Hill 1974) 239.

21. F Porz, G Grathwohl and F Thummler, Mat.Sci. and Engineering <u>71</u> (1985) 273.

22. T L Francis and R L Coble, J.Amer.Ceram.Soc. <u>51</u> 1968) 115.

23. G Grathwohl, Th. Reetz and F Thummler, Sci.Ceram. <u>11</u> (1981) 425.

24. J L Chermant, Mat.Sci. and Engineering <u>71</u> (1985) 147.

25. G G Trantina and C A Johnson, J.Amer.Ceram.Soc. <u>58</u> (1975) 344.

26. G. Grathwohl, R Hamminger, H Iwanek and F Thummler, Sci.Ceram. <u>12</u> (1984) 583.

27. K D McHenry and R E Tressler, J.Mat.Sci. <u>12</u> (1977) 1272.

28. G D Quinn and M J Slavin, Proc. 23rd ATDC meeting, Dearborn, MI (1985).

29. M G Scott, J.Mat.Sci. <u>10</u> (1975) 1527.

30. M V Swain and R H J Hanninck in 'Advances in Fracture Research' <u>4</u>, ed. D Francois (Pergamon Press 1981) 1559.

31. M Ruhle, N Claussen and A H Heuer, in 'Advances in Ceramics' <u>12</u> (Amer.Ceram.Soc. 1984) 352.

32. N Claussen in 'Advances in Ceramics' <u>12</u> (Amer.Ceram.Soc. 1984) 325.

33. N Claussen and M Ruhle, in 'Advances in Ceramics <u>3</u>' (Amer.Ceram.Soc. 1981) 137.

34. N Claussen, F Sigulinski and M Ruhle, in 'Advances in Ceramics <u>3</u> (Amer.Ceram.Soc. 1981) 164.

35. N Claussen, Mat.Sci. and Engineering <u>71</u> (1985) 23.

36. R A Sambell, A Briggs, D C Phillips and D H Bowen, J.Mat.Sci. <u>7</u> (1972) 663 and 676.

37. R A Sambell, D C Phillips and D H Bowen in Proc.Int.Conf. on Carbon Fibres' (Plastics and Polymers Conf.Suppl. No.6, 1974) 105.

38. S Yajima, K Okamura, J Hayashi and M Omori, J.Amer.Ceram.Soc. <u>59</u> (1976) 324.

39. J J Brennan in 'Tailoring Multiphase and Composite Ceramics' ed. Tressler and Bradt (Plenum 1985) in press.

40. K M Prewo in 'Tailoring Multi-phase and Composite Ceramics' ed. Tressler and Bradt (Plenum 1985) in press.

41. T N Tiegs and P F Becher, 'Tailoring Multiphase and Composite Ceramics' ed. Tressler and Bradt (Plenum 1985) in press.

42. E Fitzer and J Schlichting, in 'High-Temperature Science' <u>13</u> (1980) 149.

43. T Hirai, in 'Tailoring Multiphase and Composite Ceramics' ed. Tressler and Bradt (Plenum 1985) in press.

44. R Hayami, K Ueno, I Kondou, N Tamari and Y Toibana, in 'Tailoring Multiphase and Composite Ceramics' ed. Tressler and Bradt (Plenum 1985) in press.

Composites - the present and the future

G D SCOWEN

A new appreciation of engineering materials and their significance in product development is currently in progress worldwide, driven mainly by technical and economic pressures from many sectors, ranging from transport, construction and consumer to marine and aerospace. Fibre reinforced polymeric composite materials are exciting much of this attention and activity.

The greater the demand for product materials with a high sophistication of material content, the more attractive composite materials become for engineering developments, resulting in products with high added value. However, basic materials price stability and enhanced physical properties, compared with lightweight metallic materials, are often more 'down to earth' reasons for choosing composites.

Finally, several present and future composite product developments are reviewed with reference to their operational requirements and a composite materials solution.

The author is with the Composite Component Development Group, GKN Technology Ltd, Wolverhampton.

INTRODUCTION

Fibre reinforced plastic composite materials are exciting a great deal of interest in many industrial sectors, ranging from transport, construction, marine through to aerospace.

Over the last few decades, many of these developments were in low and medium stress environments, such as in the construction industry.

Highly stressed composite product applications have been mainly restricted to the aerospace industry. More recently, however, the automotive sector has been developing composite products which operate in highly stressed and hostile environments.

The ability to fully exploit the many attractive physical properties of composite materials for product development depends upon the availability of design and analytical techniques to model and quantify the fibre and matrix behaviour, and also to adequately quantify structural response by using finite element and advanced C.A.E. methods.

Finally, an adequate product manufacturing base is a pre-requisite to successful composite product developments.

NEW MATERIALS AND MARKET CLIMATE

The present trend is towards developing materials with increasing sophistication and information content. Composite materials fulfil this role extremely well. By the composites definition, a material can be built-up from a multiplicity of fibre, matrix and core materials to adequately fulfil many materials behavioural roles in one single material. For example, low weight, low thermal expansion, high corrosion resistance and high fatigue resistance, can be obtained using fibre reinforced plastic composites.

Figure 1 shows the heuristic trends in future material developments for engineering products. The illustration is based upon Altenpohls work, Reference 1. Each point in the figure illustrates the current degree of materials sophistication and weight of material per unit of product. The arrows further illustrate the future trends for various engineering materials. The expectation in the automobile sector is shown. The automotive sector is alive with many exciting high stress and temperature applications for polymeric

TABLE 1

SOME TYPICAL FIBRE/MATRIX CONSTITUENTS AVAILABLE

FOR COMPOSITE COMPONENT DEVELOPMENTS

Composite/ Compound	Fibre	Resin	Filler	Comments
DMC	'E' Glass	Polyester	Ca Co3	Fibre length approx 15 mm (low performance)
SMC-CLASS	'E' Glass (approx 30% by wt)	Polyester (approx 30%)	Ca Co3 Mg 0 etc	Sheet moulded. compound. CLASS C - Undirectional Continuous fibres CLASS D - Undirectional discontinuous fibre CLASS R - Random fibre (fibre length approx 50 mm)
UMC	'E' Glass	Polyester	-	Combinations of random (20%) and aligned (30%) fibres (ARMCO CORP)
XMC	'E' Glass (70% by Vol)	Polyester (approx 30%)	-	Continuous fibre orientations generally between 5° and 10° some classes have a % of random fibres

SOME HIGH PERFORMANCE COMPOSITE MATERIALS

*GLASS "PRE PREG"	Glass	Epoxide	-	CLASS E CLASS S) Higher performance CLASS R) than 'E' Glass
*CARBON "PRE PREG"	Carbon	Epoxide	-	CLASS: . High Modulus . High Strain . High Strength
*AROMATIC POLYAMIDE PRE PREG	Aramid	Epoxide	-	Denoted aramid or 'Kevlar'

EPOXIDE	
PHENOLIC	
POLYESTER	
POLYIMIDE	SOME HIGH PERFORMANCE RESINS
BISMALEIMIDE	
POLYETHER- ETHERKETONE	
POLYSULPHONE	

*All stated materials are available in continuous fibre and bulk resin form

TABLE 2

MAIN FABRICATION TECHNIQUES

.   R.R.I.M.
.   RESIN INJECTION (OR RTM)
.   VACUUM MOULDING
.   HOT PRESS MOULDING (SMC)
.   PULTRUSION (PULFORMING)
.   FILAMENT WINDING

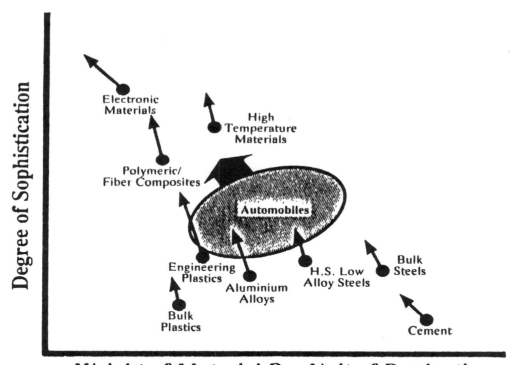

Degree of Sophistication

Weight of Material Per Unit of Production

FIGURE 1

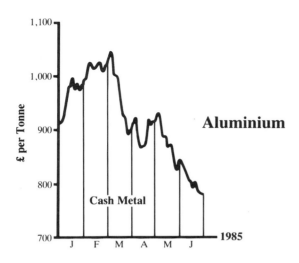

FIGURE 2

Aluminium price for a 6 month period in 1985

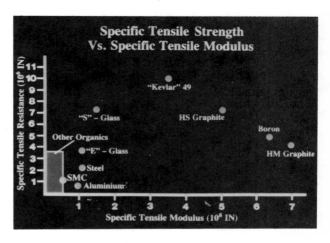

FIGURE 3

The relationship between specific tensile strength and specific tensile modulus for various composite constituents and metals

composites. Some of these developments will be described later.

Materials costs play an important part in the selection of composite materials for engineering products. Materials such as steel, glass fibre and epoxy resin have generally increased in price with inflation. High performance fibres such as carbon have gradually fallen in price, mainly reflecting an increasing demand from the aerospace sector for materials with enhanced physical properties. Carbon and glass fibre are presently less susceptible than other materials to energy cost fluctuations.

Volatility in materials price movements is illustrated in Figure 2 for a 6 month period in 1985. The price of aluminium peaked at over £1000/tonne, but only four months later a 25 per cent. fall occurred. A similar large fluctuation in the price of aluminium was experienced in early 1983, further illustrating the problems encountered in costing components manufactured in traditional lightweight materials. During a similar two year period, there was no significant price fluctuation for glass, carbon fibre or resins such as polyester and epoxide, the basic constituents of composite materials. However, no material, whether new or traditional is immune to price fluctuations, but composite materials have undoubtedly shown good price stability over recent years.

COMPOSITE CONSTITUENTS AND PROPERTIES

Table 1 shows some of the composite materials that are available to the engineer for product developments. Some high performance systems are too costly for product developments, selection often being hampered by materials processing difficulties. However, "hybridisation" allows the mixture of relatively low cost glass fibre, with more costly fibre materials, such as carbon fibre, resulting in more cost effective composite components.

The varied material properties and behavioural benefits achieved by using fibre composites in engineering products cannot be surpassed by traditional materials.

Figure 3 shows the specific tensile strength and the specific tensile modulus for various composite fibre constituents and traditional metallics. It can be seen that high modulus (HM) graphite or carbon fibre has a specific stiffness approximatley seven times greater than structural steel. Aromatic polyamide (KEVLAR - tradename) has a specific strength five times greater than structural steel. Fibrous materials can therefore, be compounded with polymeric resins into composite materials with very attractive physical properties and behavioural characteristics, thus increasing the portfolio of

engineering materials available to industry for product development.

COMPOSITE FABRICATION

Table 2 shows some of the main fabrication techniques currently used for the manufacture of composite products, the choice of the process being very dependent upon the performance and market requirements for a particular product application. The process and gelation times are generally determined by the geometrical complexity and component or product size. Typically, gel times range from a few seconds to over 90 minutes.

Compression moulding still ranks high as a technique which produces composites with high fibre volume fraction and excellent structural integrity.

Figure 4 shows an important adaption of the compression moulding technique in which a resin injection stage is used. Resin is injected at typical pressures of between 40 and 80 psi. Cycle times of few seconds to several hours for very large components, are quite common.

Figure 5 shows a large capacity, high temperature autoclave for the fabrication of large aerospace components, such as aircraft wing control flaps or helicopter rotor blades. A vacuum stage assists resin flow and removal of air or gaseous volatiles given off during the curing cycle. The pressure chamber allows a build-up of pressure on the components outer surface for compaction of the composite fabrication at chamber temperatures up to 200°C. Pressures up to 3000 psi are quite common, and composite components fabricated by this method, have excellent surface finish and structural integrity.

Figure 6 shows a large universal helical filament winding machine fabricating a large g.r.p. storage tank. Fibre rovings (with resin) are wound onto a large cylindrical former or mandrel. Fabrication equipment of this large scale, illustrates the wide range of composite products possible and also the large resources commitment that must be made in plant and factory space.

Numerically controlled gantry machines for "pre-preg" tape laydown are also used for fabricating large aircraft wing sections. Robots are being used for fabric cutting and "pick and place" operations for composite materials and pre-form handling.

COMPOSITE PRODUCTS

It is a difficult task to review the many exciting composite product developments that

FIGURE 4

Large bulk seed hopper fabricated by resin
injection and hydraulic press equipment

FIGURE 5

Large capacity, high temperature autoclave
for fabricating composite aerospace components

FIGURE 6

Universal helical filament winding machine -
tank for chemicals under fabrication

FIGURE 7

Composites in the construction sector -
grp roof and panels

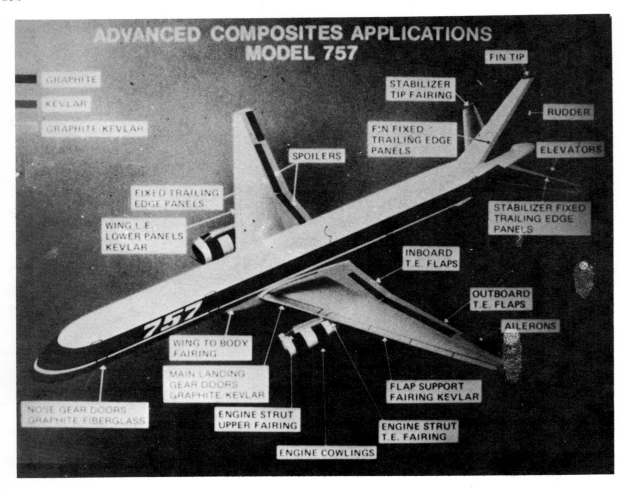

FIGURE 8

Advanced composites applications model 757

FIGURE 9

B1 Horizontal stabiliser - Rockwell-Grumman

FIGURE 10

Westland Combat Helicopter

are currently taking place in the following market sectors:

- Electrical and electronics
- Corrosion resistant market
- Construction
- Appliance and business equipment
- Consumer products
- Marine
- Land transport
- Aerospace

It has been estimated that there will be a global demand approximately 2½ million metric tonnes for fibre reinforced plastic composites during 1985. By 1995 it has been estimated that approximately 4 million metric tonnes will be used for product developments, a very attractive and healthy growth in demand prospect for the composite materials industry.

CONSTRUCTION.

The construction industry gradually introduced fibre reinforced plastic composites into build-ing constructions in the 1950's. Pipe work for chemical plant, where stress corrosion was a primary failure mode, was fabricated using filament wound glass reinforced polyester resin.

Water tanks for high rise flats, septic and sewage treatment tanks also came within this corrosion resistant market. The traditional construction markets for reinforced plastic are building panels for walls, glazing, facia materials, roofing and roof support beams. Figure 7 illustrates some of these composite products in a building which was designed to have low maintenance. The domed glass reinforced plastic (grp) roof feature, reduces self-weight and therefore, reduces the overall support structure loads, leading to a relatively lower construction cost compared to one in concrete.

The benefits of using composite materials in the construction industry are:

- Environmental stability and durability
- Physical properties benefits;
  i.e. high stiffness,
       high strength,
       low weight and corrosion resistance
- Low maintenance
- Ease of handling
- Design flexibility and aesthetics

AEROSPACE

Figure 8 illustrates the many composite developments taking place for civil aircraft products. Although polymeric composites have been used for some years for civil aircraft cabin interior fittings, the application of composites to highly stressed components, such as the fuselage and the wing has not occurred.

The Boeing 757 programme identified potential composite components, such as edge members, spoilers, flap assemblies, cowlings etc.

The Grumman composite nacelle or engine cowling development, Reference 2, for the DC6 represent a major application of composite materials in a civil aircraft. However, many of the major composites development have been for combat aircraft, such as the BA Jaguar aircraft engine bay door and the Rockwell B1 horizontal stabiliser shown in Figure 9, probably the largest composite primary structure ever built, was less costly and weighed 500lb less than a comparable metal stabiliser.

Helicopter applications, Figure 10, have yielded some interesting fibre reinforced plastic and honeycomb sandwich composite constructions, Figure 11. The Westland Helicopter composite rotor, shown in Figure 12, illustrate a "BERP" rotor tip profile geometry that gives increased aerodynamic lift. This particular design detail in conventional metal, would be more costly to manufacture.

Some of the advantages in using composite rotor blades are:

- Designed to a specific weight, the distribution of weight can therefore be controlled more accurately.
- Improved control of blade profile, hence increased aerodynamic efficiency
- Improved damage tolerance over metals
- Benign failure mode
- Infinite fatigue life. Three or four blade changes are often required for metallic rotors.

Therefore, the rotor composite construction clearly illustrates the importance of a material with a high information content.

Although the constituent materials for the composite are more costly than metals, the compression moulding manufacturing method is less costly, and therefore, the composite rotor blades are generally cost competitive with metal blades.

Figure 13 shows a rather unusual application of composites to the AD500 airship front pod section of the gondola, extensively fabricated in aromatic polyamide fibre reinforced plastic composite.

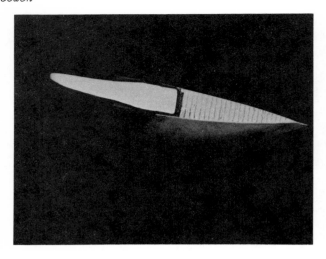

FIGURE 11

Fibre reinforced plastic and honeycomb
sandwich composite main rotor.

FIGURE 12

Composite "BERP" rotor tip for increased
aerodynamic efficiency

FIGURE 13

The AD500 Airship - front pod section
fabricated in aramid fibre composite

FIGURE 14

The HMS Brecon-Hunt class mine countermeasures vessel

FIGURE 15

The LR4 North Sea submersible - pressure hull in grp

FIGURE 17

Composite Sports Products

FIGURE 16

MCV80 armoured personnel carrier

FIGURE 18

Canoes fabricated from aramid or polyester fibres

MARINE

Figure 14 illustrates probably one of the largest marine applications of grp composite. The HMS Brecon was the first of the hunt-class mine countermeasures vessels, her plastic hull was reinforced with 160,000 square metres of woven fabric.

Figure 15 illustrates the Vickers-Slingsby, LR4 North Sea submersible. The grp hull resists the very high crushing and compression pressure forces that are exerted at depth.

prendre photocopie + ajouter quelques avantages

DEFENCE

Figure 16 illustrates the GKN MCV80 military combat vehicle, in which fibre reinforced composite is used extensively for inner armour and serves as protection for the occupants.

CONSUMER

Figure 17 and 18 shows the important consumer market sector for fibre reinforced composites. Tennis, squash, badminton rackets, skis, fishing rods, sailing craft such as canoes are available in carbon, glass or aramid reinforced plastic composite. Lightweight quality of finish, high stiffness and strength as well as corrosion resistance are all properties that promote the use of composites in leisure goods.

MEDICAL

Figure 19 illustrates the medical application of composites. The CT scanner is currently installed in many hospitals, and consists of an X-ray source. Since the carbon fibre reinforced plastic composite couch is transparent to X-rays, a more accurate diagnosis results with reduced dosage rates to the patient undergoing treatment.

The last two sectors to be described in this presentation are automotives and space technology. These industries will produce some of the most exciting applications of composite materials in future product developments.

AUTOMOTIVES

The automotive sector has for many decades, been one of the main employers of labour and producers of manufactured goods in this country. The world recession has hit the motor manufacturers hard and to compete there is currently a drive towards improved productivity and improved fuel or energy economy in road vehicles. Fibre reinforced plastic composites will, therefore, contribute to improving the performance of road vehicles, now and in the future.

Over the last 20 years, there have been many composite product developments for lightly or moderately stressed automotive components. However, composite components for highly stressed applications have been limited to the low volume specialised car market. Often, the large capital investment required for fabrication plant is prohibitive for many companies and can only be embarked upon when component supplier works closely with the automotive manufacturer.

Fibre reinforced composite were first used by Ford some fifty years ago. The car consisted of 14 wood fibre reinforced phenolic vacuum moulded composite panels, bolted into a steel structural frame, resulting in a vehicle 30 per cent. lighter than a similar sized car of that era.

In 1940, Owens Corning built an experimental car in body panels fabricated in glass fibre reinforced polyester resin, probably the first such application of this composite materials in an engineering structure.

The General Motors Corvette illustrated in Figure 20 was first introduced in 1953 as a low volume, two seater sports car, with glass fibre reinforced polyester body panels. This vehicle formed the basis of many exciting new composite developments for automotives, especially in highly stressed applications. For example, the introduction in 1981, of a glass reinforced composite leaf spring, Figure 21, with 80 per cent weight saving compared with a steel spring and a high energy absorbing face bar bumper.

Figure 22 illustrates the Lotus Eclat Excel low volume sport car. The body shells are manufactured by moulding a top and bottom pair of shells. The manufacturing process consists of a vacuum assisted resin injection technique using glass reinforced polyester. There is provision at the moulding stage to incorporate the paint primer prior to the application of the gel coat. However, using a similar paint finishing technique, it should be possible to eventually mould-in the final paint finish.

The box beam and sandwich construction is formed by wrapping uni-directional glass fibre around low density polyurethane foam cores.

During the resin injection cycle, the resin flows around the cores to produce high performance fibre reinforced beams for withstanding side collision etc. Figure 23 shows Esprit body shells assembled and under test. Although the resin injection cycle and demould times can take up to 2 hours, the potential to mould-in many components such as windows, general accessories, as well as the final paint finish, brings into focus the flexibility in design and manufacture that is possible with composite materials components.

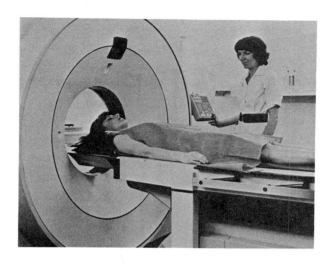

FIGURE 19

CT scanner with integral couch moulded using carbon fibre.

General Motors 1980's Corvette

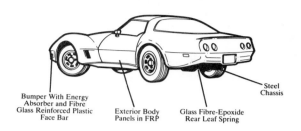

Bumper With Energy Absorber and Fibre Glass Reinforced Plastic Face Bar

Exterior Body Panels in FRP

Glass Fibre-Epoxide Rear Leaf Spring

Steel Chassis

FIGURE 20

General Motors 1980's Corvette

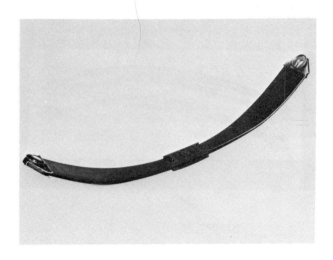

FIGURE 21

Corvette grp composite leaf spring

Lotus Eclat Excel

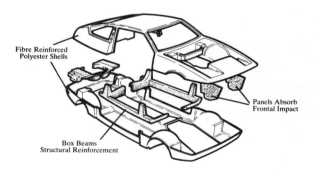

Fibre Reinforced Polyester Shells

Panels Absorb Frontal Impact

Box Beams Structural Reinforcement

FIGURE 22

Lotus Eclat Excel

FIGURE 23

Lotus Esprit on Test

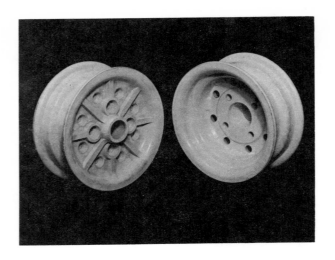

FIGURE 24

Composite automotive wheels

FIGURE 25

Composite wheel during road testing

FIGURE 26

GKN Composite Leaf Spring

Fibre reinforced composite wheels have been under active development over the last decade. Figure 24 shows some typical compression moulded wheels. Companies such a Pontiac and Goodyear are planning the low volume production of 100,000 units annually for a sports car application in 1987. The wheels are composed of mixtures of glass reinforced resin, along with quantities of carbon fibre in some applications which are supposedly cost/competitive with steel wheels. The wheels offer the benefits of improved strength, stiffness and running characteristics, accompanied by greater styling freedom. Weight saving is between 30 and 50 per cent compared with steel wheels. Moulded wheels are "true round" and therefore, wheel balancing is eliminated. The Goodyear wheel has logged more than 4 million miles on vehicle road tests through arduous terrain.

Figure 25 shows a Mini Metro composite wheel being put through its paces during road test trials at N.E.L., on behalf of an industrial consortium, Reference 3.

The GKN Composite Leaf Spring, shown in Figure 26, represents probably the first high volume application of a polymeric composite in a highly stressed and hostile environment. The spring units are manufactured in glass reinforced plastic with a possible weight saving of at least 50 per cent over the equivalent steel leaf springs. This particular development emphasises the large research, development and manufacturing resources required to put a cost/effective product using new materials into the high volume automotive market place. The extensive testing programme utilised special purpose multi-axis test rigs which applied simultaneously a combination of vertical loads, lateral loads, roll and wind-up, based data acquired from vehicles with instrumental suspension systems. The composite springs were subjected to slalom reversals, single wheel and two kerb strikes, corrugations, pave, pave, snatch starts and pot hole braking. Validations were also carried out in typical automotive environments, such as extremes of successfully carried out in typical automotive environments, such as extremes of temperature, moisture and corrosive chemicals.

The NEL sulcated spring concept is shown in Figure 27 and is representative of a second generation suspension concept which is not purely limited by simply replacing the existing metal spring. The design concept is based on the full exploitation, by using innovative design, and the unique material properties of fibre composite.

The geometry has several advantages over more traditional spring solutions:-

. Spring stiffness can be tailored in an x-y-z plane. This effectively, if required, allows a stiff transverse location for the suspension member with an accompanying vertical spring compliance, effectively a "compliant strut".

. Fatigue strength of the component can be maximised since the composite material is being used in bending.

. The load/deflection can be linear or variable rate, by simply varying material thickness, width or material fibre orientation.

Figure 28 shows a composite automotive suspension bracket or "wishbone". The component development consisted of the design and manufacture of a composite suspension bracket, using glass fibre reinforced vinylester composite with a quantity of directional fibres. The composite component was fabricated using a compression moulding technique in one single stage.

Figure 28 also shows the moulded-in web configuration which gave additional stiffness to the suspension arm structure. The design aims were:

. To achieve the correct compliance for specified ride characteristics

. To carry the required suspension loads

. To tailor and optimise the composite materials lay-up configuration; this was generally dictated by the manufacturing process

Although the composite suspension bracket had a similar weight to the original metal "wishbone", it was thought that additional weight saving would result in a future development, since the mounting bush assembly could be virtually eliminated. The benefits derived from using composite materials were a corrosion resistant suspension member, and reductions in noise levels and vibration.

The Bertin composite suspension, shown in Figure 29, was developed from a complete car design project. The vehicle weighing 650 kg, would have a fuel consumption of 125 miles per gallon at speeds of 56 mph. The study identified that 60 per cent of the empty weight of the vehicle was spread equally between the structure, the power system and the suspension. The study showed that weight saving in the suspension element would be extremely difficult. Bertin's solution was to integrate several suspension functions in just three major composite components, using a suspension spar. The spar copes with the vertical suspension loads and the horizontal loads due to drag and braking. The conventional control arms and drag strut or anti-roll bar combinations necessary in conventional suspensions were, therefore, eliminated.

The suspension spar was manufactured in a glass fibre reinforced composite by a filament

FIGURE 27

The N.E.L. sulcated spring

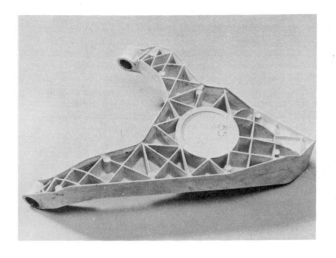

FIGURE 28

Composite suspension arm with moulded webs

FIGURE 29

The Birtin suspension

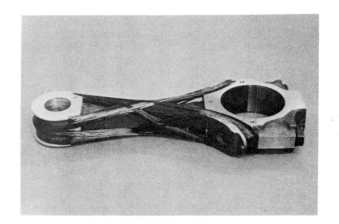

FIGURE 30

The Fibre Reinforced Composite connecting rod

A Composite Concept Engine Block and Crankcase

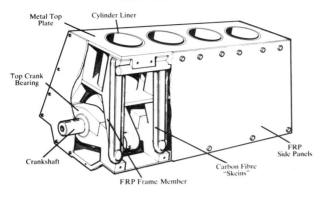

Metal Top Plate
Cylinder Liner
Top Crank Bearing
Crankshaft
FRP Frame Member
Carbon Fibre "Skeins"
FRP Side Panels

FIGURE 31

A composite concept engine block and crankcase

FIGURE 32

The Shuttle Orbiter. Composite doors, engine pods, arm booms etc.

FIGURE 33

Main support structure for space telescope - Intelsat

winding and pressing route. The direct weight reduction over a conventional Macpherson strut arrangement was specified as being greater than 50 per cent.

High temperature composite applications for automotive components is probably one of the most exciting and demanding uses of new materials with great potential for commercial exploitation and the development of components with high added value.

There has been a recent trend in conrod design and manufacture, to move away from the traditional forged steel components. Aluminium has been a candidate replacement material, but more recently fibre reinforced composites have been developed in Europe and the USA for connecting rods. Figure 30 shows a carbon fibre reinforced composite connecting rod developed at I.K.V. (Aachan).

The component consists of a central filament wound stem which locates both the gudgeon pin bearing with the crankshaft bearing. The bearing section is of metallic form. High temperature composite component applications, such as the I.K.V. development, demonstrate that high temperature polymers can be operated at temperatures between 100° and 200°C.

Some of the benefits of using low weight composite conrods were:

. Reduced inertia forces

. Less bearing wear

. Smaller bearings

. Crankshafts with reduced mass

. Reduced noise

The final automotive product development to be discussed in this paper is the composite engine.

There have been recently several developments to produce low weight fibre reinforced automotive I.C. engines in Europe and the USA, Polymotor in the USA, and Harwell in the UK, are two contenders in this particular composite development. Figure 31 shows a concept composite engine that Harwell have recently developed, Reference 4. The engine is assembled from basic fibre reinforced plastic composite frame members that are secured to bearers supporting the crank shaft by tying or looping skeins of filament wound carbon fibre. The side and end panels are also fabricated in fibre reinforced plastics, and a metal plate containing the cylinder liners is attached to main frame members.

Some of the main benefits derived from using a composite engine block are:

. Low weight

. Reduced noise levels

. Improved thermal efficiency

. Possibility of in-mould incorporation of many design features

SPACE TECHNOLOGY

This industrial sector will undoubtedly, yield many new and exciting composite material developments. The design requirements for lightweight and dimensional stability over temperature ranges of between -250° and +250°F, can be adequately met with composite materials. Figure 32 illustrates the space shuttle orbiters maneouvering engine pods and arm booms fabricated in carbon fibre reinforced plastic composite.

Figure 33, illustrates the structural complexity of a main support structure fabricated in carbon fibre composite for an Intelsat telescope.

CONCLUSION

The choice and selection of a particular new material for component development depends on the interaction of a wide range of variables, such as:-

. Market conditions and material price.

. Material properties and behaviour, as well as design manufacture, testing and quality assurance.

The engineer and management must not be prejudiced by their traditional or historical view of materials.

Composites have outstanding potential for building in a high information content, thus enabling products to be developed with a high added value.

The potential of composite materials was reviewed with reference to some exciting present and future composite component developments.

REFERENCES

1. ALTENPOHL, D.G. Materials and Society, 1979, 3, 315.

2. ANDERSON, R and POVEROMO, L.M. Composite Nacelle development. Proc. 29th Ann. Conf., The Society of the Plastics Industry, 1984, pp 11-C-L - 11-C-5. New York: The Society of the Plastics Industry.

3. WOOTON, A.J., HEDRY, J.C. CRUDEN, A.K. and HUGHES, J.D. Structural automotive components in fibre reinforced plastics. Paisley-Col. of Tech./SDA/NEL/3rd Int. Conf. on Composite Structures, Paisley, 1985.

4. AERE Harwell OX., UK Patent No. GB2140502

Mechanical alloying - the development of strong alloys

M J FLEETWOOD

Mechanical alloying is a solid-state process for making alloys by high-energy milling, under conditions such that constituent powders are repeatedly fractured and welded together and ever more intimately mixed. After subsequent consolidation at elevated temperature, the alloys can be shaped by rolling, forging and machining. The process is used to incorporate a fine dispersion of ceramic particles. MA nickel-base superalloys, combining a dispersion of yttria with conventional precipitation strengthening, have achieved higher strength at 900–1100°C than directionally solidified and single-crystal alloys, and are being used for gas-turbine vanes and blades. MA ferritic stainless steel, with outstanding strength and corrosion resistance at temperatures as high as 1300°C, has been produced as sheet, tube, plate, rings and forging forgings. MA aluminium alloys also offer higher strength, for example in as-forged thick sections of Al–Mg–Li alloy.

The author is Manager of the New Business Development in the Technology Centre of Inco Alloy Products Ltd, Birmingham.

INTRODUCTION

The development of higher strength capability in metallic materials has generally been achieved by increasing the number and level of alloying additions. The sequential development of wrought iron, carbon steel, low alloy steels and highly alloyed steels is an obvious example and there are parallels in copper and aluminium alloys. But perhaps the most celebrated sequence of alloy development is that of the nickel superalloys, and it was in seeking to raise their strength limits by another notch that mechanical alloying was developed.

Mechanical alloying originated from a long search for the means to add high-temperature strength conferred by a fine dispersion of ceramic particles to intermediate-temperature strength developed by conventional alloying. From the initial laboratory success[1] in 1968 the process has been developed into a well controlled production operation, whole series of nickel, iron, aluminium and other alloys have been designed specifically to use the process, and techniques have been developed to form and fabricate the alloys into useful components.

After 7 years of development, MA materials are now in use in a range of applications, but work continues to extend their use.

In this paper, the place of mechanical alloying in nickel-base superalloy development is outlined, before describing the process and the results of applying it to the development of nickel, ferrous, and aluminium alloys of improved strength.

Although the intention is to show how further strength improvement has been achieved in alloys already working near their limits, reference is also made to improved corrosion resistance and other properties offered by MA materials.

STRENGTHENING OF NICKEL-BASE SUPERALLOYS

The development of superalloys has been dominated by the incentive to increase operating temperature of gas turbines. Like all heat engines, the efficiency of the gas-turbine engine increases with the temperature of the inlet gas, and the development of more efficient and powerful engines has required alloys able to operate at increasingly high temperatures. The greatest demand for high-temperature strength is made for the first-row blades and vanes in the high-pressure turbine, for they see the hottest gas. Such blades and vanes are made from Ni-Cr and Co-Cr base alloys, partly because of good resistance to high temperature oxidation and corrosion, but also because the nickel chromium base, especially, tolerates the addition of high levels of alloying

Table 1 Compositions of Commercially Produced MA Nickel- and Iron-base Alloys

Alloy	Nominal Composition (weight %)											
	C	Cr	Al	Ti	Ta	Mo	W	Zr	B	Y_2O_3	Fe	Ni
INCONEL Alloy MA754	0.05	20	0.3	0.5						0.6		bal.
INCONEL Alloy MA6000	0.05	15	4.5	2.5	2	2	4	0.15	0.01	1.1		bal.
INCOLOY Alloy MA956		20	4.5	0.4						0.5	bal.	

Table 2 Stress-Rupture Properties of INCOLOY Alloy MA956 Sheet

Temperature °C	Stress (MPa) to Produce Rupture in		
	10h	100h	1000h
Longitudinal			
982	84	75	67
1100	64	57	51
1150	56	50	44
1200	53	47	42
Transverse			
982	76	68	61
1100	62	56	50
1150	36	33	29
1200	32	29	25

Table 3 Compositions of MA Aluminium Alloys

Alloy	Nominal Composition (weight %)					
	C*	O*	Mg	Cu	Li	Al
IN-9052	1.1	0.8	4			bal.
IN-9021	1.1	0.8	1.5	4		bal.
MA Al-Li	1.1	0.4	4		1.5	bal.

*present as carbide and oxide

Table 4

Typical Properties of Mechanically Alloyed and Conventional Aluminium Alloys of Similar Composition

Alloy:	IN-9052 (1)	5086-H34	IN-9021 -T4/T6 (2)	2024-T4	MA Al-Mg-Li (1)	7075-T73
density, Mg/m^3	2.66	2.66	2.78	2.78	2.57	2.52
dynamic elastic modulus, GPa	74.4	74.4	76.5	75.1	80.6	
0.2%yield stress, MPa	390	254	500/560	324	430	410
UTS, MPa	470	324	570/600	468	500	480
elongation, %	13	10	12	19	10	7
fracture toughness K_{IC}, MPa m$^{1/2}$	46		30/44	32	30	33

(1) as forged

(2) forged, solution treated (500°C), cold water quenched, aged naturally (T4) or at 135°C (T6)

Rupture in 100 hours at stress of 138 MPa

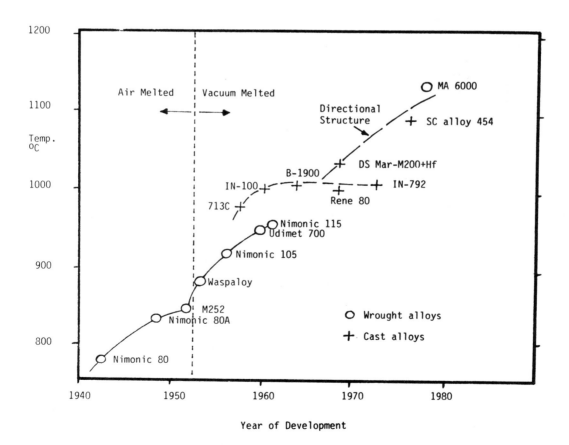

Figure 1: Progressive development of stress-rupture temperature capability in superalloys

elements without forming brittle intermetallics and can therefore be strengthened to high degree.

Solid-Solution and Precipitation Strengthening

Strength is developed in superalloys by introducing heterogeneities into the structure which successfully impede the movement of dislocations, even at high temperatures. The simplest impediments are solute atoms: at temperatures up to 0.6 Tm (815°C), the most potent solid-solution hardeners in nickel are aluminium, tungsten, molybdenum and chromium, while in the creep range (> 0.6Tm) the slow diffusing elements, molybdenum and tungsten, should be the most effective hardeners.[2] Solid-solution strengthening is used in nearly all superalloys, but by itself it is insufficient to provide high strength. The other heterogeneities commonly used to impart strength are particles fine enough to interact with dislocations: in nickel superalloys precipitation of a second phase is the main source of high strength, particularly at high temperature. Carbides have been widely used to strengthen grain boundaries, but the most important matrix strengthener is precipitation of the γ' intermetallic phase based on $Ni_3(Al,Ti)$.

Nickel-base superalloys have reached their important position as high-temperature materials largely through the remarkable characteristics of the γ' precipitate. Its constituents may be completely soluble in the austenitic matrix at 1100°C, yet form a fine precipitate constituting more than 50% of the alloy at 900°C. With the same fcc structure and almost the same lattice parameter as the matrix, γ' nucleates homogeneously, has a low surface energy and exhibits long-time stability. Its strength actually increases with temperature up to 650°C and yet it is ductile in massive form.

The strengthening effect of the γ' precipitate depends principally on particle size and distribution.[2] Mitchell[3] has shown that hardness typically increases with particle diameter, then falls. On the ascending curve, γ' is cut by dislocations which glide in pairs; the resulting anti-phase boundary hardening increases with volume percent of precipitate, size of γ' and APB energy. On the descending part of the hardness curve, dislocations by pass γ' particles by looping or by dislocation climb; again creep strength is increased by γ' volume fraction. Because γ' grows at a significant rate at temperatures > 0.6Tm, long-time creep strength is greatly affected by stability of the γ'. Generally, stability is improved by minimising the lattice parameter mismatch between γ and γ', and by reducing the solubility and coefficient of diffusion of γ' forming elements in the matrix.

The influence of alloy composition on these various factors is complex[4], but principal γ'-forming alloying additions are aluminium, titanium, niobium and tantalum. Carbide-forming elements are added to nickel-base superalloys principally to precipitate carbides on grain boundaries and hence increase high-temperature creep resistance by inhibiting grain-boundary sliding. Again, the influence of alloy composition is complex[4], but carbides are formed mostly by chromium, molybdenum, titanium, tantalum and niobium and modified by minor additions of boron and zirconium.

Limits to High Temperature Strength

Using the alloying additions and strengthening mechanisms outlined above, many different nickel-base superalloys have been developed in the last 40 years, to reach higher levels of stress-rupture and creep strength. Figure 1 illustrates the increase in temperature capability of aircraft gas-turbine blade alloys achieved by successive waves of alloy development. γ' strengthened wrought alloys reached their limit with NIMONIC* Alloy 115 and UDIMET 700, in which the volume fraction of γ' was increased to the extent that the practical temperature range of hot workability, the gap between the γ'-solvus and the liquidus, became very narrow. Any further increase in γ' resulted in unworkable alloys, and subsequent, stronger compositions had all to be made as castings.

Cast alloys were again improved by further γ' and carbide strengthening, until the level of alloying additions threatened alloy stability, by forming sigma, and other brittle phases of topologically close-packed structure, during long-time service. These phases are formed through electron bonding of electropositive elements, such as chromium, molybdenum and tungsten, with electronegative elements, including nickel, cobalt and iron. Strength of conventionally cast superalloys was pushed to the limit by precise control of composition, working close to the sigma-phase boundary, and by limiting the content of the electropositive elements. Reduction of chromium content required the use of reliable coatings to compensate for loss of high-temperature oxidation and sulphidation resistance.

The strength attainable in the best conventionally cast alloys was then limited by the relatively weak grain boundaries transverse to the stress direction, and in order to raise creep strength even higher, it was necessary to remove transverse grain boundaries. That was achieved by developing production techniques for directional solidification of castings and, later, for the growth of single crystal castings. Those techniques have taken alloys hardened by carbides and γ' precipitation to their limits of strength; they are faced with the inherent limitation that precipitated phases coarsen and dissolve as the melting point is approached.

Dispersion Strengthening

One means of strengthening at high temperature not used in the alloys so far discussed is dispersion hardening, by a fine dispersion of refractory oxide, nitride or carbide. Because the dispersoid

---

*Trade Mark of the Inco family of companies

is a ceramic of very high melting point, it can be stable right up to the melting point of the alloy in which it is dispersed. It has not proved possible to distribute particles fine enough to strengthen the alloy uniformly through the structure by incorporating them in the melt, so all successful attempts to dispersion strengthen superalloys have used a powder metallurgy route.

Nickel was successfully dispersion strengthened by thoria, introduced by mechanically mixing nickel and thoria powders, hot compacting and extruding.[5] Thoria-dispersion-strengthened Ni and NiCr were also made using nickel precipitated chemically onto thoria particles.[6] Both alloys proved to have better stress-rupture strength than conventional superalloys at temperatures above about 1000°C, but at lower temperatures the absence of precipitation strengthening made them comparatively weak. To achieve improved temperature capability without sacrificing strength at intermediate temperatures, it was obviously necessary to combine dispersion stengthening with γ' precipitation hardening.

The methods by which the thoriated alloys were made cannot be used for alloys containing more reactive elements such as aluminium and titanium. Chemical routes lead to oxidation of the reactive elements, while mechanical mixing to produce a fine dispersion requires ultrafine powders, which, for reactive elements, are highly pyrophoric and difficult to handle. A new process was required.

THE MECHANICAL ALLOYING PROCESS

High-Energy Milling

Mechanical alloying makes possible the combination of dispersion, solid-solution and precipitation strengthening by mechanically mixing all the constituents in powder form ever more intimately until diffusion completes the formation of a true alloy powder.[7] The mixing is achieved by dry high-energy ball milling, under conditions such that powders are not only fragmented but also rewelded together - a process prevented in conventional ball milling by use of liquids and surfactants.

Throughout the period in which mechanical alloying was being developed, the high-energy milling was performed in attritors, in which the ball charge is stirred vigorously with rotating paddles. First commercial production used attritors able to process up to 34kg of powder per charge, but, in the latest production units, up to 1 tonne of powder is processed in a 2m-diameter mill containing in excess of a million balls, weighing some 10 tonnes.

The charge is a blend of elemental and prealloyed powders, at least one of which is a ductile material, and crushed master alloy, containing intermetallic compounds of the most reactive elements. For example, titanium and aluminium are added as Ni-Ti-Al alloy, which has much lower activity than do elemental titanium and aluminium.

To provide a dispersed phase in nickel and iron-based alloys, fine inert oxides can be included in the charge, usually Y_2O_3.

Figure 2 illustrates the effect of a single high-energy collision between two balls on powder trapped between them. The ductile elemental metal powders are flattened and, where they overlap, the atomically clean surfaces just created weld together, building up layers of composite powder, between which are trapped fragments of the brittle powders and the dispersoid. At the same time, work-hardened elemental or composite powders fracture. These competing processes of cold welding and fracture occur repeatedly throughout the milling, gradually kneading the composites so their structure is continually refined and homogenised. After the initial stage of milling (Figure 3a), the composites show coarse layers of identifiable starting materials, with the dispersoid closely spaced along the welds.

After more refinement through fracture and rewelding, the composites develop a structure of convoluted lamellae (Figure 3b) of decreasing thickness between welded surfaces, along which dispersoids are closely spaced. The combination of severe cold work and heating from the kinetic energy of the balls aids diffusion, and as diffusion distances continually decrease by finer mixing of the constituents, solute elements dissolve, areas of solid solution grow in the composite powders and metastable phases may precipitate.

In the final stage of milling (Figure 3c), the lamellae become more convoluted and thinner (1μm or less) and the composition of individual particles converges to the overall composition of the starting powder blend. Precipitation of equilibrium phases occurs, work hardening and softening reach a balance and the micro-hardness of individual powder particles attains a saturation value - around 650kg/mm^2 for FeCr alloys.[7] Finally, the lamellae are no longer resolvable optically, being less than 0.7μm thick, and the distance between dispersoid particles along the weld interfaces approximately equals the spacing between the welds. Then, the composition of individual powder particles is equivalent to that of the starting blend and mechanical alloying is complete: further milling will produce no further homogenisation in either the matrix or the dispersion.

Mechanical alloying is not simply mixing on a fine scale: true alloying occurs. The progress of alloying has been monitored by X-ray diffraction of powder samples after varying times of processing: in both aluminium alloys[8] and steel,[9] diffraction peaks for alloying elements had disappeared on completion of mechanical alloying. Also, by completion of the process any contaminant present in the charge or picked up from the mill will have been refined and distributed uniformly through the structure, so there are no large inclusions which might subsequently act as stress raisers and crack initiators.

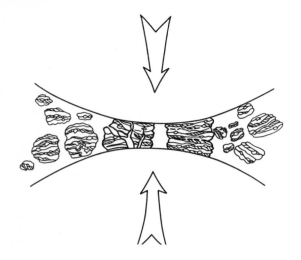

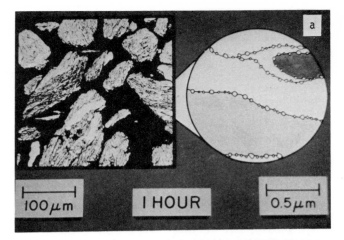

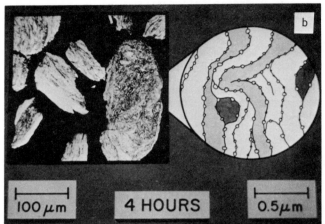

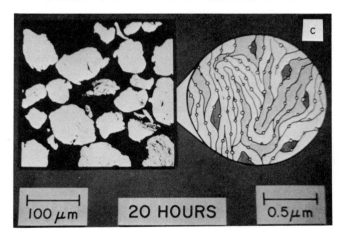

Figure 2: Effect of a single collision between two balls on trapped powder

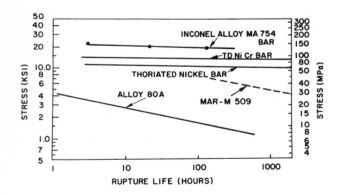

Figure 4: Stress-rupture properties at 1093°C of INCONEL MA754 compared to other bar materials

Figure 3: Stages in the mechanical alloying process

(a) Early Stage

(b) Intermediate Stage

(c) Final Stage

Consolidation and Structure Control

Oxide-dispersion-strengthened powders do not densify on simple sintering and the high hardness of MA powder prevents cold pressing, so the powders discharged from the mill must be consolidated by a combination of high temperature and pressure - hot compaction + hot extrusion, hot extrusion in a can, or hot isostatic pressing. Extrusion of canned powder is generally preferred, as being cheapest and able to provide anisotropic structures with good properties. High-temperature soaking prior to hot consolidation completes homogenisation by diffusion.

Any further processing after extrusion must be carefully controlled to generate the optimum grain structure for particular applications. Two structures are beneficial:

(i) fine, equiaxed grains, giving best room-temperature strength, fatigue strength and workability (e.g. for subsequent forging)

(ii) coarse, elongated grains, giving highest high-temperature stress-rupture strength and thermal-fatigue resistance.

The fine-grain condition exists in extruded material, whereas coarse, elongated grains are developed by secondary recrystallisation of material thermomechanically processed to produce a high level of stored strain energy. It is a complex matter to control the stored energy, which depends on the whole thermomechanical processing history of the material: from the conditions of milling, through the temperature, ratio and speed of extrusion, to the details of post-extrusion processing, including hot and cold rolling or forging, and the grain coarsening anneal. Achievement of the highest strength attainable by an MA material demands the closest control of production.

NICKEL-BASE MA SUPERALLOYS

Two nickel-base superalloys are now produced commercially, INCONEL* Alloys MA754 and MA6000: compositions are shown in Table 1. INCONEL ALLOY MA754 has a solid-solution strengthened Ni-20%Cr matrix with only low levels of titanium and aluminium, and contains about 1 volume % of Y_2O_3 dispersoid. INCONEL Alloy MA6000 is a critically balanced composition, giving a high level of γ' strengthening combined with dispersion strengthening, and good high-temperature oxidation and sulphidation resistance. γ' strengthening is provided by high levels of aluminium, titanium and tantalum and dispersion strengthening by 2.5 volume % of Y_2O_3. Tungsten and molybdenum provide solid-solution strengthening and the balance of chromium and aluminium with titanium, tantalum and tungsten combines oxidation resistance with good sulphidation resistance.

MA754: Production and Properties

Nickel and chromium are charged to the ball mill as, respectively, 4-7μm and -150μm elemental powders, aluminium and titanium are charged as -150μm crushed master alloy, and Y_2O_3 is added as 1μm agglomerates of 20-40nm particles. On completion of mechanical alloying, powder is canned, extruded and hot rolled to bar, or hot and cold rolled to sheet, under conditions leaving high enough stored strain energy to grow large, elongated grains on annealing.

Stress-rupture properties of MA754 **recrystallised** bar tested at 1093°C in the longitudinal direction are compared in Figure 4 with those of TD-Ni, TD-NiCr, NIMONIC* Alloy 80A and the cast cobalt alloy MAR-M 509. Like other dispersion-strengthened materials, MA754 has a very flat log stress-log life curve and has notably high strength for rupture in long times. Compared with a 100 hour rupture stress of 102 MPa at 1093°C in the longitudinal direction, the rupture stress in the transverse direction is only 42 MPa, reflecting the important effect of the elongated grain structure. Recrystallised bar exhibits a strong (100) texture in the longitudinal direction which results in a low elastic modulus (151 GPa) and good thermal-fatigue resistance.

MA754: Forming and Application

Nozzle guide vanes, machined from recrystallised extruded and hot-rolled bar have been used very satisfactorily in the GE F404-100 engine used to power the US Navy's F18 'Hornet' fighter.[10] With the aim of reducing cost, forging to near net shape has been developed[11] to the production stage. Fine-grain extruded bar has been conventionally coined, upset and forged, or forged direct to vane and blade shapes, (Figure 5) which can be successfully recrystallised to a coarse, elongated grain structure with the required (100) texture. Forgings for fuel nozzles have also been produced in MA754. Since forging adds to the total thermomechanical processing, structure can be optimised by choice of forging conditions: a high forging temperature has been used to achieve a 100 hour rupture strength at 1100°C of around 112 MPa.

MA754 has also been formed by hot spinning and ring rolling.[11] Although grain structures differ appreciably from those of recrystallised bar, good tensile and stress-rupture strengths have been achieved and further work aims to optimise the use of such forming routes.

MA6000: Processing and Properties

Nickel is charged to the ball mill as 4-7μm elemental powder, chromium, molybdenum, tungsten and tantalum as -150μm elemental powders, and the most reactive elements aluminium, titanium, boron and zirconium as -150μm crushed nickel-base master

---

*Trade Mark of the Inco family of companies

*Trade Mark of the Inco family of companies

alloys. Y_2O_3 is added as 1um agglomerates of 20-40 nm particles: on completion of mechanical alloying the dispersoid is a mixed oxide of Y_2O_3-Al_2O_3

After consolidation by canning and extrusion at 1050°C, bar is hot rolled up to 90% reduction in thickness and recrystallised by zone annealing to form coarse, elongated grains, many of which extend the length of the bar.[12] To optimise properties, the recrystallised alloy is heat treated 0.5h/1230°C, AC to take γ' into solution and 2h/955°C,AC + 24h/845°C,AC to precipitate γ' of best size and distribution.

Figure 6 compares the specific rupture strengths (strength/density) for 1000 life as a function of temperature for MA6000, directionally solidified (DS) MAR-M 200+Hf and single-crystal alloy PWA 454. At intermediate temperatures around 760°C, MA6000 approaches DS MAR-M 200+Hf in strength and is progressively stronger than the cast alloys at temperatures above about 900°C, as γ' strengthening declines. To maintain strength at high temperature in the face of environmental attack, MA6000 is less reliant on coatings than other alloys: MA6000 combines the cyclic oxidation resistance of Alloy 713C with the sulphidation resistance of Alloy IN-738.[13]

MA6000 seems particularly suited to small blades operated uncooled at high temperature: blades machined from bar have been engine tested. Forging to near net shape is economically desirable, but, with its high-temperature strength, the alloy cracks at chilled surfaces when conventionally forged. Alternatives being investigated are: isothermal forging at low strain rates using heated dies, and conventional forging at high strain rates with thermal barriers to reduce surface chilling.[11] Fully recrystallised blades with grains of high aspect ratio have been made by both routes and the effect of forging conditions on structure and properties is being studied.

MA STEELS

Various types of ferrous alloy have been made by mechanical alloying, including 17%Cr, 7%Ni, 1.2%Al precipitation-hardened austenitic martensitic steel[9] and Fe-25Cr-6Al-2Y[14] However, the most highly developed material is the 20%Cr, 4.5%Al ferritic steel INCOLOY* alloy MA956, dispersion strengthened with 0.5% Y_2O_3 (see Table 1 for composition); MA956 has been made in the form of bar, sheet, plate, wire, tubing, forgings, rings, hot spinnings and fabrications.

MA956: Production and Properties

MA956 is mechanically alloyed from elemental iron powder, FeCrAlTi master alloy crushed to -150μm and Y_2O_3 powder, and hot compacted by extrusion. Bar is made by extrusion and hot rolling followed

---

*Trade Mark of the Inco family of companies

by static annealing to attain the recrystallised grain structure, elongated along the rolling direction. Plate and sheet require hot and cold rolling, with hot cross rolling to obtain 'pancake' shaped grains after recrystallisation, giving isotropic properties in the plane of the sheet. The first step in tubemaking is hot compaction by direct upsetting against a die in the extrusion press, to provide a 300 mm diameter billet. That is bored to a hollow preform, conventionally extruded to tubeshells, and tube reduced on a Pilger mill. The resulting fine-grained tube is recrystallised to coarse grains elongated along the tube axis.

The stress-rupture strength of MA956 bar in the longitudinal direction at 1000°C is more than twice those of alloys such as INCOLOY Alloy 800H, INCONEL Alloy 617, cast HP alloy modified with Nb and HP-50WZ.[15] In sheet form, MA956 also exhibits high stress-rupture strength (Table 2), ten times that of HASTELLOY X at 1093°C, while 0.2% yield strength is 201 MPa at 600°C, 97 MPa at 1000°C and still 76 MPa at 1200°C[16]

The high-strength capability is combined with exceptional high-temperature oxidation and corrosion resistance, associated with formation of an alumina oxide scale.[15] In cyclic oxidation resistance, MA956 is superior to HASTELLOY X, INCONEL Alloy 601, INCOLOY Alloy 800, HK steel and 310 stainless steel. The alumina scale is an excellent barrier to carbon and no carburisation occurs in hydrogen-methane mixtures at 1000°C. Sulphidation resistance is also good.

MA956: Forming, Fabrication and Application

MA956 is basically a ferritic stainless steel and has a ductile to brittle transition in the temperature range 0-80°C, dependent on product form and stress rate.[16] Therefore, forming should be performed at a temperature of at least 80°C. The alloy is readily bendable and can be deep drawn and stretch formed; spinning is best done hot (400-800°C) to avoid springback and the alloy hot die forms similarly to high-strength titanium alloys.[16] A variety of rings have been rolled up to 1m internal diameter, with circumferentially elongated recrystallised grain structures and properties equivalent to those of hot-rolled bar.[11] Spun dish shapes have shown even better strength than rolled rings.[11]

Like all dispersion-strengthened materials, fusion welding agglomerates dispersoid distribution and disturbs the grain structure, reducing strength in the weld zone.[14] Fusion welds should be confined to joints outside the hottest sections and for gas sealing. Flanged, threaded, rivetted and brazed joints have all been used successfully.

MA956 was originally developed for use in sheet form in gas-turbine combustors, but, with its combination of high strength up to 1300°C, corrosion resistance and formability, the alloy has found many other applications. Gas-turbine applications under development include: fabricated nozzles, inlet plenum and compressor nozzle parts

Figure 5: MA754 gas-turbine components conventionally forged from extruded bar

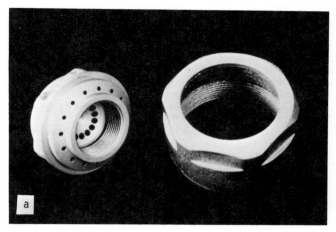

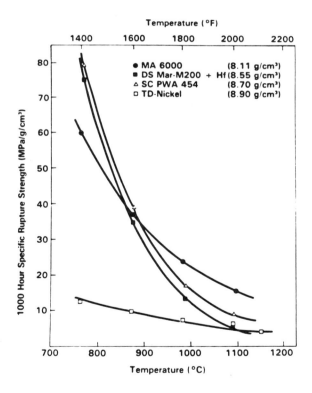

Figure 6: Specific stress-rupture strength for 1000-hour life of INCONEL Alloy MA6000 compared to other high-strength alloys

Figure 7: Power-station combustion components in INCOLOY Alloy MA956

(a) Burner nozzles machined from forgings

(b) Air-stream swirlers fabricated from sheet

of vehicle turbines; rings for aeroengine combustors, and a combustor baffle for industrial turbines. Use in power stations include oil and coal burners and swirlers, and fabricated tube assemblies for fluid-bed combustion.[11] Examples of power-station burner nozzles and swirlers are shown in Figure 7.

MA ALUMINIUM ALLOYS

Mechanical Alloying of aluminium alloys results in a dispersion of carbides and oxides, which not only strengthens the alloy directly but stabilises a fine grain structure, adding further to strength. Because strength originates from the dispersoids, composition of the alloy matrix can be designed principally for corrosion resistance and toughness rather than strength. Alloying elements added to conventional aluminium alloys for precipitation strengthening and for grain size control, such as manganese and chromium in 5086, can be omitted.

On these principles, three alloys have been developed to date (Table 3). IN-9052 is the equivalent of a 5000 Series alloy, requiring no heat treatment and offering good strength in thick sections; IN-9021 is heat treatable by solution treatment + natural or elevated-temperature ageing and is the equivalent of a 2000 series alloy. MA Al-Mg-Li offers inherent high strength in thick section, combined with low density.

Production

Mechanical alloying of aluminium alloys is currently conducted in attritors, with a capacity (for scale-up and evaluation programmes) of 50 tonnes/year.[17] Processing starts with attriting of elemental powders with the organic process control agent, such as stearic acid, needed to bring cold welding into balance with powder fracture. No dispersoid is added to the charge because oxide on the surface of the starting powders and process control agent deposited on the powder surface are incorporated into the composite powders during mechanical alloying as hydrated oxides and carbonates. Repeated fracture and welding lead eventually to a fine dispersion of 10-30nm particles throughout the structure. On completion of mechanical alloying the alloy has a heavily cold-worked, dynamically recrystallised structure with grains as fine as 0.05μm[18].

Because the dispersed hydrated oxides and process control agent liberate hydrogen and nitrogen on heating the alloy, the powder is vacuum degassed at a temperature higher than the maximum temperature to which the alloy will be subjected during subsequent processing and service. This treatment at elevated temperature completes homogenisation of the matrix, reduces carbonates to Al_4C_3 (which form most of the dispersoid), and causes grain growth to around 0.1μm.

The powder can be compacted by hot compaction or hot isostatic pressing, but vacuum hot pressing has been favoured, as giving minimum grain coarsening. 130kg billets 300mm diameter are made in a cylindrical die, with the end plate pushed into the powder by a ram, operating through a vacuum seal in the furnace lid. Scale-up to 500 kg, 470mm diameter billets is possible on this vacuum hot press. The billets are conventionally extruded to 110 mm diameter and below, and forgings are made from extruded stock. Recrystallisation and grain growth on extrusion result in grains of about 0.3μm, which are resistant to further growth because grain boundaries are pinned by the dispersoid. This fine grain size is an order of magnitude smaller than that observed in other high-strength powder-metallurgy aluminium alloys and is believed to be partly responsible for the high strength of the MA materials[18]

All the MA aluminium alloys can be readily formed from extruded stock by either hammer or press forging.

Properties

Typical mechanical properties are summarised in Table 4 in comparison with those of conventional alloys of similar density. IN-9052 and the MA Al-Mg-Li alloy develop their high levels of strength as-forged and the alloys therefore have intrinsic thick-section capability, without the problem of heat-treatment distortion. The heat treatment of IN-9021 is chosen to suit the shape of finished component and the properties required. In extruded material, the longitudinal properties can be up to a yield strength of 580 MPa, UTS of 620 MPa, elongation 12%. In all the alloys, higher tensile properties can be achieved, at the expense of toughness, by varying the process conditions and, in IN-9021, by varying heat treatment. IN-9021 is substantially stronger in fatigue than T175-T736 and 7050-T736 forgings and both IN-9021 and IN-9052 are more resistant to corrosion than would be expected of the base composition[19] IN-9052 can also be stressed to near the yield point in stress-corrosion environments without cracking.

With their combination of high strength and corrosion resistance, the MA aluminium alloys are being developed for aerospace and other applications.

CONCLUSION

The process of Mechanical Alloying can be applied to many alloy systems to make all manner of materials. However, the milling step takes appreciable time, carefully controlled thermomechanical processing is frequently needed, and the cost would not be justified to make materials for which there is already an established production route. In fact, mechanical alloying is no more a run-of-the-mill powder metallurgy process than is single-crystal casting a process to be found in every foundry. It is a process for making alloys from normally incompatible components - immiscible liquids, metastable phases, incongruent-melting intermetallics, cermets. It is beyond the scope of this paper to describe application to superconductors, corrosion-and wear-resistant coatings, supercorroding alloys and the like[7], but to date much the most important application of the process has been to the dispersion-strengthened alloys described above.

Production of mechanically alloyed powder is now well over 100 tonnes/year and the process is making a major contribution towards pushing metallic materials to their limits, indeed to temperatures previously thought to be the preserve of ceramics. Perhaps the refractory dispersoid makes MA materials at high temperature as much like a ceramic as a ductile metal can be.

Acknowledgement

The author is indebted to Inco Alloy Products Ltd. for permission to publish this paper.

References

(1) J.S. Benjamin: Met.Trans. 1, 1970, 2943-2951

(2) R.F. Decker: Symposium on Steel Strengthening Mechanisims, Zurich, 5-6 May 1969.

(3) W.I. Mitchell: Z. Metallkunde, 57, 1966, 586

(4) R.F. Decker and M.J. Fleetwood: Metals Society Conference on Production and Applications of Less-Common Metals, Hangzhou, 8-11 Nov. 1982.

(5) D.K. Worn and S.F. Marton: Powder Metallurgy, Interscience Publishers, New York, 1961, 309-340

(6) G.B. Alexander: 1965, French Patent No: 1,424,902

(7) P.S. Gilman and J.S. Benjamin: Annual Rev.Mat.Sci., 13, 1983.

(8) J.S. Benjamin and R.D. Schelleng: Met. Trans.A, 12A No.10, Oct. 1981, 1827-1832

(9) P.Pant and H. Grewe: Tech. Mitt. Krupp Forschungsber. 38 (2), Aug.1980, 103-121

(10) Inco's New ODS Superalloys in the 80's: Metal Powder Report, Jan 1983, 32-35

(11) E. Grundy, C.J. Precious, D. Pinder: Metal Powder Report PM Aerospace Materials Conference, Berne, 12-14 Nov. 1984

(12) R.L. Cains, L.R. Curwick, J.S. Benjamin: Met. Trans 6A, Jan 1975, 179-188

(13) Y.G. Kim, H.F. Merrick: NASA CR 159 493 (Contract NAS3-20093), Lewis Research Centre, May 1979

(14) F.G. Wilson, C.D. Desforges: Behaviour of High-Temperature Alloys in Aggressive Environments, Proc. Petten International Conference, 1979, Paper 63.

(15) R.H. Kane, E.M. McColvin, T.J. Kelly, J.M. Davidson: NACE Corrosion Meeting, New Orleans, April 1984.

(16) IncoMAP Data Sheet: INCOLOY alloy MA956

(17) R.D. Schelleng, S.J. Donachie: Metal Powder Report, 38(10), Oct.1983, 575-576

(18) S.J. Kang: Proceedings ASM Conference, Philadelphia, Oct.2-6, 1983

(19) P.J. Bridges: Metal Powder Report PM Aerospace Materials Conference, Berne, 12-14 Nov., 1984

Aluminium alloys for airframes: limitations and developments

C J PEEL

Aluminium alloys have been the predominant choice for material used in the construction of civil and military aircraft since World War Two. They are, however, subject to competition from other materials, particularly titanium and carbon fibre reinforced composite. This paper illustrates some of the developments that have occurred with conventional alloys and defines the limitations to further exploitation in terms of availability, cost, structural efficiency and compatibility with existing manufacturing technology. Some new prospects for aluminium-based materials are considered, including conventional alloys, metal matrix composites and powder alloys.

The author is with the Materials and Structures Department, Royal Aircraft Establishment, Farnborough, Hampshire.

INTRODUCTION

Aluminium alloys have been used in ever increasing amounts in the construction of the airframes of both civil and military aircraft since the Second World War. Considerable improvements have been achieved during these four decades in terms of both engineering properties and, possibly more importantly, in terms of understanding the limitations that restrict their more adventurous use. 'Since aluminium alloys are the predominant choice for the material in virtually all aircraft structure, improvements in their performance are perhaps more of benefit to the aircraft manufacturer than to the aluminium industry. Indeed, it could be argued that developments in aircraft structural materials bring more disadvantages to the manufacturer in terms of the cost of establishing the safe limits for the

use of the new developments than the improvements often justify. The aircraft manufacturer tends to be conservative in his choice of materials and unfortunately as the understanding of any material's characteristics increases so does the number of the complex tests that needs to be undertaken to prove the safe limits for the application of that material in any new or modified structure. For example, fatigue failure proved itself to be a major limitation to the exploitation of aluminium alloys during the 1950's and 1960's to the extent that every new airframe and significant modification to an airframe now has to undergo a very expensive major fatigue test. It is intended to expound the limitations imposed by fatigue problems in more detail.

Since aluminium alloys are currently the predominant choice in the limited selection of aircraft structural materials available, it is clear that their utilisation is hardly hindered by technical difficulties and there is little incentive for performance improvement especially if any extra risk or extra cost were to be introduced in the implementation of the development. However, whilst aluminium alloys currently dominate the scene they are not without competition from other metallic materials, particularly titanium alloys, and from composite materials. It can be seen [Fig 1] that the distribution of material in past military and civil airframes has favoured the aluminium alloys but that brand new designs of the 1990's could well include a significant use of composite material. It is emphasised that the distribution for the date 1990 is based only on the most futuristic projections for a few of the most advanced aircraft. Most of the aircraft currently in the design stage are not able to exploit composite to such an extent especially in the case of the larger structures. Nevertheless, the potential competition from new composite and metallic systems, including superplastic forming of titanium alloys, has led to a more aggressive evaluation of the further potential for improvement of aluminium despite the cost

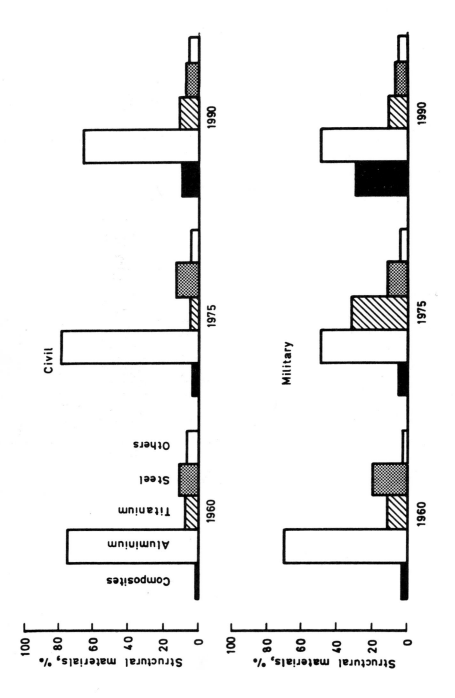

Fig 1. The approximate distributions of
materials used in existing military and civil
aircraft and those predicted to be selected
for new aircraft of the 1990's.

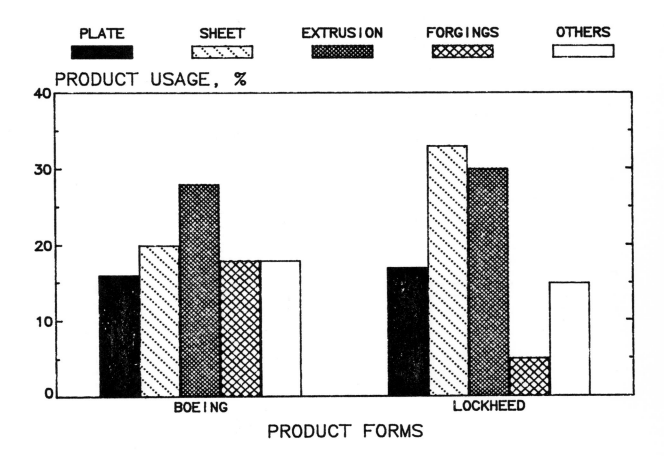

Fig 2. The approximate distributions of
aluminium alloy product forms used in large
transport aircraft (1,2).

Peel

penalties and the increased technical risks involved.

It is worth considering the advantages of the competing materials to analyse those areas in which aluminium alloys could be considered deficient. Areas critical to the choice of material are considered by the author to be availability, cost, structural efficiency and compatibility with existing manufacturing techniques. The limits imposed on the use of aluminium alloys in these categories will now be considered partly in relation to the performance of the competing materials.

MATERIALS AVAILABILITY

A characteristic of aluminium alloys in present usage is that they have been produced in the main by semi-continuous casting techniques capable of producing alloy ingots in the range of 5 to 10 tonnes, possibly even 20 tonnes depending upon the factory and the alloy considered. A notional 10 tonnes of aluminium alloy has a volume of 3.5 cubic metres sufficient to produce a piece of rolled plate 2 metres wide, 100 millimetres thick, and possibly 17 metres in length, ie a piece large enough to enable the production of the wing cover of a transport aircraft essentially machined from a single very large plate. Not only could the size of the cast ingot be critical in this respect but limitations may well arise from the size of the heat treatment facilities or the stretching machine since virtually all aluminium alloy plate used in aircraft structure is stretched after quenching for the relief of residual stress. Similar arguments can be raised for the other product forms, for example in the length of stretched extrusions and the width of cold rolled sheet. Very large investments have been made by the aircraft manufacturers in equipment capable of machining very large pieces basically because the elimination of joints is both financially and technically attractive. It is commonly the case that the presence of joints in load bearing structure introduces both fatigue and corrosion problems.

A conventional metallic aircraft structure requires the constituent aluminium alloys in a range of product forms[1,2]. A large transport aircraft is likely to require roughly equal proportions of plate, sheet and extrusion, with extrusions perhaps surprisingly being the largest single category chosen in the construction of some US aircraft [Fig 2]. The type of aircraft, its size and the manufacturer's equipment are, of course important factors. A smaller quantity of forged material would be required amounting to perhaps 5% of the structure mass. Ideally, to achieve a significant penetration of the aircraft structure a new aluminium alloy or material would need to be available in more than one of these product forms. This applies especially to some of the more advanced developments now beginning to appear such as the metal matrix composites. That is, to produce large pieces of structure, either the integral machining of large pieces of

stock material or the building-up of structure is required from a mixture of sheet and extrusions for example. An advanced material with an exceptionally high stiffness available in extruded form may prove to be incompatible with conventional sheet to which it must be attached. Similarly, there may prove to be considerable difficulties in producing carbon composite covers on internal metallic structure. To summarise, aluminium alloys need to be available in a complete range of product forms and, for many applications, in as large a size of product as equipment will allow.

COST

Without any doubt cost is a critical factor in the implementation of a developing material. The cost of implementation of a new material can be considered to be the combination of basic material costs and those accrued in its application. Aluminium alloys are normally considered to be cheap material and a price of £1-£2 per pound avoirdupois is frequently quoted for semi-finished product. This compares favourably with competing titanium alloys and composites which are at least ten times as expensive. However, the rate of utilisation of aluminium is low, that is sometimes only 10% of the metal purchased is built into the aircraft. The very nature of the construction of a composite component leads to a high utilisation and the difficulties in machining and forming the titanium alloys encourage the use of close-to-form products with high utilisation. It becomes clear that utilisation levels are important to the cost effectiveness of aluminium alloys as is the ability to reclaim and reprocess scrap alloy in both primary and secondary circuits. It has already been stated that the major aircraft manufacturers are equipped for the integral machining of large pieces of structure from thick plate with only limited possibilities for an increase in utilisation rates. However, a new trend is emerging in the manufacture of built-up structure in the automated computerised cutting and forming of sheet material. In some cases it would appear that this automated control can be extended to heat treatment and to mechanical assembly. The full impact of automation technology on the application of aluminium alloys has yet to be realised but it will clearly have a marked effect upon costs and possibly even on structural mass.

A further important trend is the increasing use of close-to-form products, in particular precision die forgings and extrusions for which the utilisation of material can be much higher than the previously quoted 10%. It is envisaged that the emerging very expensive metal matrix composites will prove most cost effective in these product forms. At present these specially produced materials (typically the combination of conventional aluminium alloy with a ceramic additive in particulate, whisker or fibre form) can be two orders of magnitude more expensive than conventional alloys [Fig 3].

52

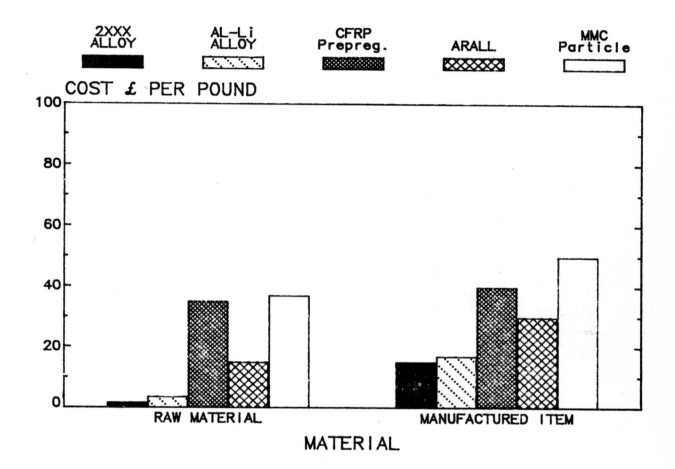

Fig 3. The estimated relative costs of conventional alloys and new materials as raw material and in built up structure.

Fig 4. Superplastically formed parts
produced in the new aluminium-lithium alloy
8090. Courtesy of BAe Warton Division.

Whilst the cost of the stock metal is of significance to the cost of an airframe, the manufacturing costs may be as much as ten times more significant. That is a die forging emerging from the press may cost £300 as forged and £3000 after machining, heat-treatment, final machining, protection, painting and final fitting. Thus the cost of a new material may be diluted by a large amount by the added value of the manufacturing process. It is important then that any new material does not increase the cost of manufacture. A possible development that may even reduce manufacturing cost is SQUEEZE CASTING, in which cast aluminium alloys are solidified under pressure to combine the cheap effectiveness of the casting process with the greater integrity of the forging.

A further major development beginning to emerge is the application of SUPERPLASTIC FORMING of selected aluminium alloys. This process allows a manufacturer to produce a complicated shape from a single formed piece of sheet avoiding expensive joints, multiple forming operations, and the need to build-up detailed structural components, thereby saving both cost and possibly structural mass. At present an adequate selection of aluminium alloys has proved to be superplastic. This includes SUPRAL alloys based on the aluminium-copper system, 7475 based upon the aluminium-zinc system, 8090 based on the aluminium-lithium system and certain of the powder metallurgy metal matrix composites. It has been established[3] that to control the problem of cavitation during superplastic forming of aluminium alloys a superimposed hydrostatic pressure should be used. This discovery will undoubtedly relax some of the restrictions imposed in the use of superplastically formed aluminium, although a considerable effort is still required to fully establish the mechanical properties of metal after forming. Limitations still arise however because the presses required in the process limit the technique to relatively small components perhaps no more than one metre square. It is well known that titanium alloys have an advantage over aluminium at present in their ability not only to be superplastically formed but also simultaneously to be DIFFUSION BONDED. The strongly adherent oxide coating on aluminium alloys renders diffusion bonding difficult for aluminium alloys, but not impossible. The combination of both techniques enables the production of more complicated and larger pieces of structure but introduces a host of new questions in terms of the structural integrity of components manufactured by these techniques. Some superplastically formed parts are shown [Fig 4], these were produced in the new 8090 aluminium lithium alloy.

Any extra cost of a new material must be related to the value of any benefits achieved in its application. To an aircraft engineer this may mean a saving in structural mass or an increase in performance within a particular envelope. The value of any mass saving depends upon the application. For example, an automobile manufacturer may be prepared to pay a premium of a few pence per pound of metal to save a pound of structure weight, the manufacturer of a civil airliner will pay possibly two or three pounds sterling per pound of metal to save the same amount and the engineer building a space vehicle will accept a premium of several hundred pounds. For civil transport requirements cost effectiveness really depends upon fuel price, whilst for military requirements performance requirements may dominate. It is clear from the previous statements that manufacturing costs dominate the cost of aluminium alloy structures with potential savings in the areas of utilisation levels, automation, and superplastic forming. The more expensive materials such as carbon fibre composite and titanium alloy will have to show significant savings in manufacturing costs to offset their high initial price. In this respect composite materials show some promise although manufacturing techniques are currently highly labour intensive. The very expensive metal matrix composites of the future will need to demonstrate large mass savings to justify their application on a significant scale, although limited application to critical parts may be more attractive.

To summarise cost is a serious limitation to the development of any new material. With conventional aluminium alloys basic material costs are low but even so a modest increase of times two for a new raw material would appear to require a 10% mass saving to justify its application in transport aircraft. Similarly an equivalent increase in manufacturing costs accrued in applying a new material of between 5% and 10% would require a mass saving of 10% to be cost effective. A plot of estimated structural mass savings and possible costs is included [Fig 5].

STRUCTURAL EFFICIENCY

Superficially it might appear that technical developments for aluminium based materials are virtually exhausted, but it has been explained that because aluminium alloys are "used in anger" developments have been limited to relatively small improvements often aimed at safer application. It is worth considering the structural efficiency of current aluminium alloys and future developments. This is done in a simplified way by considering the weights of unit areas of panels, stressed either in tension [Fig 6] or compression [Fig 7]. The limits imposed in structural efficiency for structure loaded predominantly in tension are those of minimum gauge, strength and fatigue strength.

Minimum gauge requirements stem from an inability to handle material that is of less than a certain thickness and furthermore because of the vulnerability of very thin material to damage. Aluminium alloys behave relatively well in this respect and a typical minimum gauge for thin sheet would be 0.6 mm, although honeycomb skinning may even be as thin as 0.3 mm. Composite materials perform relatively badly because a certain number of plies is required to achieve a modicum of isotropy and with the resin based matrices the uniformity and integrity of thin composite skins is questionable. In practice, therefore, a minimum gauge for CFRP material is likely to be approximately twice that for aluminium.

Strength levels are readily available for a range of alloys, in the present examples designs have been assumed to be critical in terms of ultimate strength rather than in inelastic stability and minimum values of ultimate strength have been used for comparison purposes. Comparative ultimate design allowables for CFRP have been based upon a limiting strain of 0.4%. Fatigue limitations are more difficult to specify because, although fatigue strengths of aluminium alloys are well established, the fatigue stresses experienced by the metal vary considerably throughout a structure. Thus a limiting strain in a transport aircraft bottom wing skin may be of the order of 0.45% to achieve a life of 100,000 flights but the limiting hoop strain in the pressure cabin may be only 0.15%. Equivalent cut-off strains for fatigue critical areas of the shorter-lifed military strike aircraft are still unlikely to exceed 0.5%, effectively limiting aluminium alloys to stress levels little more than half of the potential strength obtainable. Currently the factors limiting CFRP to the relatively low strains quoted are variability, notch sensitivity and environmental degradation. It has been assumed that the fatigue performance of the CFRP at this strain level will prove adequate but this still needs to be resolved.

The relative tensile performance of the selected materials is plotted [Fig 6] showing aluminium alloys to perform reasonably efficiently at low loading indices but to be severely handicapped by the cut-off fatigue limit at higher loads. The high strength aluminium alloys would be attractive if static strength alone were the criterion. Since the fatigue requirements effectively limit the strains allowed in tension structures it has become normal practice to use the lower strength 2000 series alloys for these areas because although weaker than the 7000 series alloys they possess slightly better fatigue properties. It seems possible that the 8000 series aluminium-lithium alloys may be similarly exploited because they appear to demonstrate a further improvement in fatigue performance with reduced density as an added bonus.

Recognising the severe limitations imposed on the use of aluminium by fatigue, Vogelsang[4] and his colleagues have been developing a composite material based on aluminium skins covering plies of aramid fibre reinforced epoxy. This ARALL material achieves improved fatigue performance by pretensioning the fibres to put the aluminium skins into compression. At the same time there are further benefits of increased strength and reduced density. However, such a material is not easy to produce, especially in thin gauges, and early examples demonstrated inadequate residual strength once the fibres were cut. There remain the problems of formability, anisotropy, decay in pretensioning with service, and relatively reduced compression performance. However the concept of selectively reinforcing aluminium alloys for fatigue critical applications may prove to be a significant step forward in the production of lighter fatigue-resistant structures. There are indications that weight savings of as much as 30% could be achieved for fatigue-critical structures using ARALL.

Compressive performance requires a different balance of properties. In the present example the performance limits selected for the comparison are minimum gauge, buckling resistance and compressive strength. The weights of long flat panels that will not yield, buckle or break are plotted [Fig 7] for a high strength 7000 series alloy, the new high strength aluminium-lithium alloy 8091, and a CFRP laminate of 50% aligned fibres and 50% at ±45° to the compressive axis. It can be seen that the performance of the aluminium alloys is competitive in the minimum gauge regime and when compressive strength is critical but that the composite gains when elastic stability is required because of the very high level of specific modulus[5]. The loading index levels chosen are typical of the ranges experienced in the lower wings of a variety of military and civil aircraft. Just as ARALL reinforced material may prove of benefit in tension loaded structures, so it may prove possible to employ fibre reinforced metal matrix composite material for applications where elastic stability is critical in compression. Relatively rigid fibres would be required, probably SiC. The metal matrix composite would have a significant advantage over its non-metallic predecessor in its distinctly higher transverse properties derived from the metallic matrix.

A consistent feature of all these simplistic calculations of structural efficiencies has been a proportional dependence of the structural mass on the density of the selected material. The new aluminium-lithium alloys achieve a significant 10% density reduction by the addition of approximately 10 at % lithium[6]. Further density reductions may prove possible in future alloys although these may not be amenable to production by ingot metallurgy. Provided that there is no reduction in alloy properties a density reduction is by far the most effective way to produce a saving in structural mass. In the author's opinion much of the potential benefit of the current CFRP material is in its reduced density compared to aluminium alloys placing it on a par with magnesium alloys but with improved specific stiffness and making it suitable for the manufacture of lightweight covers and other thin structures.

The structural performance detailed above has been based upon the assumption that there is no particular requirement for performance at excessively elevated temperature. However, there are a large number of potential

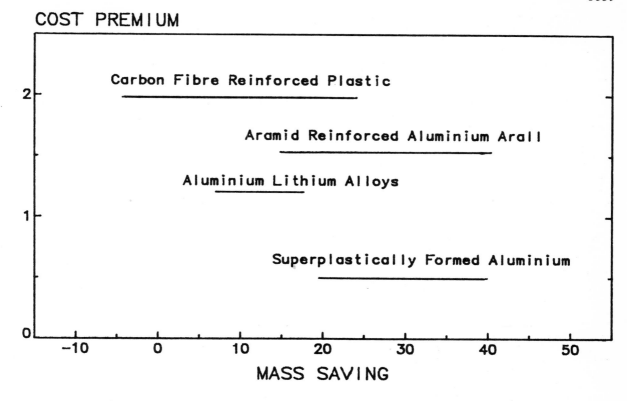

Fig 5. Potential mass savings for new materials and techniques with their predicted costs compared to conventional aluminium structure.

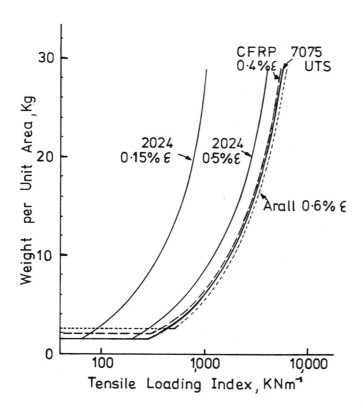

Fig 6. Predicted weights of panels of selected materials that will produce the required fatigue life or tensile strength.

57

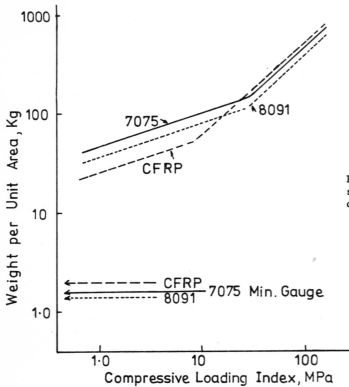

Fig 7. Predicted weights of panels of selected materials that will not yield, buckle or break under increasing compressive loads.

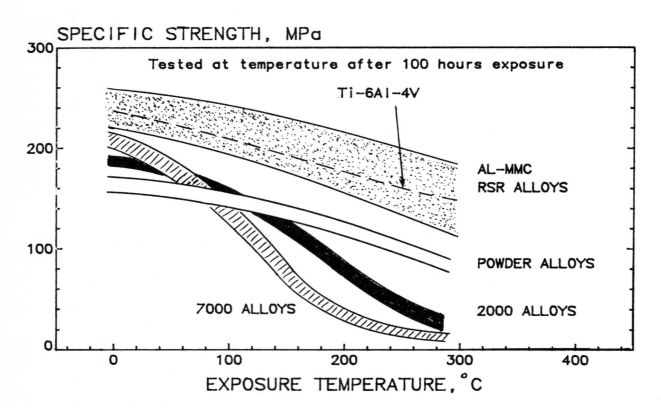

Fig 8. Elevated temperature performance of selected conventional alloys and new materials, plotted as specific tensile strength.

applications for aluminium alloys in future airframes weapons and engines that may well require an improved high temperature performance. The relative specific strengths of a selection of alloys are compared with the performance of the common Ti-6Al-4V alloy as a function of increasing temperature [Fig 8]. It can be seen that the conventional aluminium alloys become deficient at sustained operating temperatures much in excess of 160°C. There seems little prospect for significant improvement in alloy systems employing conventional age hardening mechanisms because precipitation readily occurs at temperatures above this cut-off level. There does seem to be considerable scope for the application of powder based aluminium alloys reinforced with ceramic additions, dispersion hardened with oxides or carbides, or even produced by rapid solidification technology with supersaturations of elements of low diffusivity, such as iron.

Other features that limit the structural efficiency of aluminium alloys occur with problems such as stress corrosion cracking. It has long been understood that very high strengths may be obtained from conventional aluminium alloys particularly in the 7000 series but that these high strength levels tend to be associated with low levels of resistance to stress corrosion cracking. Moreover, to efficiently use a high strength, high working stresses are required exacerbating any cracking problems. Developments in the heat treatment practices of the high strength alloys such as the double ageing treatments have resulted in a better balance between strength and resistance to stress corrosion and its associated problems but with a strength level reduced to no more than 500 MPa (0.2% proof stress). Recently progress has been achieved in the development[7] of rapid high temperature ageing treatments that impart high stress corrosion resistance with little loss in strength to the selected 7000 series alloys. The applicability of this technique of RETROGRESSIVE AGEING to a range of alloys and products needs to be evaluated but it does appear possible to produce 7000 series alloys capable of operating at very high strength levels with a minimal risk of stress corrosion cracking. The comparative performance of these improved heat treatments is adequately illustrated [Fig 9]. As a simple rule, an adequate level of stress corrosion performance would be achieved in an alloy possessing a threshold strength of more than half its 0.2% strength.

It can be seen that the comparison of stress corrosion performances [Fig 9] includes some of the new powder alloys and mechanically alloyed materials that show promise but as yet have to find significant applications. These materials produce very high stress corrosion resistance at high strength levels essentially because they do not develop the pronounced layered grain structures typical of conventional wrought products. It is well established that stress corrosion and the

related exfoliation corrosion occurs by the attack of the layered short transverse grain boundaries of conventional aluminium alloys.

A further corrosion-related limitation to the efficient use of aluminium alloys is the requirement for thin sheets of the alloys to be clad with nearly pure aluminium for corrosion resistance. Many aircraft fuselage skins employ ALCLAD aluminium alloys, and whilst the appearance of the clad materials is undoubtedly better, and the use of cladding perhaps unavoidable for polished skins, it introduces new limitations. Firstly, the strength of clad material is reduced from that of the bare core by an amount approaching 10%, secondly the fatigue strength of the alloy is further reduced by the presence of the soft skin and finally under special crevice conditions the cladding can be corroded preferentially at an alarming rate producing a form of corrosion that undermines paint schemes and adhesive bonds. The use of cladding introduces a further operation in the production of sheet and produces difficulties in heat treatment, when times at high temperature have to be controlled to prevent diffusion of copper from the alloy core into the clad skin. Problems occur during aircraft manufacture in achieving good surfaces for adherence of adhesive used in many structural applications and in the case of chemically milled sheet in ensuring that the relative thicknesses of core material and cladding are correct after milling from one side. However, cladding does impart a much improved corrosion resistance for material in store and for manufactured components waiting in the factory for finishing.

The use of anodising and chromate-containing primers is now well established as a means of achieving good corrosion resistance even in the absence of cladding and is successfully employed in many painted aircraft skins. However, these protective schemes do not solve the problems of corrosion in the factory environment and of polished skins. Moreover, anodising will itself reduce fatigue strength [Fig 10] especially when applied to clad material by an amount dependent upon the type of film ie whether chromic acid, phosphoric acid or sulphuric acid and upon the extent to which the anodised surface is subsequently sealed. The use of chromic acid anodising and chromates in inhibiting paint schemes also give cause for some concern for environmental reasons.

It is not the purpose of this paper to review the corrosion behaviour of aluminium alloys but it should be indicated that in terms of the cost of maintaining aircraft in service for as long as 30 years corrosion is the major problem adding very significantly to the costs of the aircraft operators[8]. There is as yet little information available as to the longevity of the composite materials intended for structural applications and as to the extent that their properties may be degraded by exposure to real environments. It is of course well established that the performance

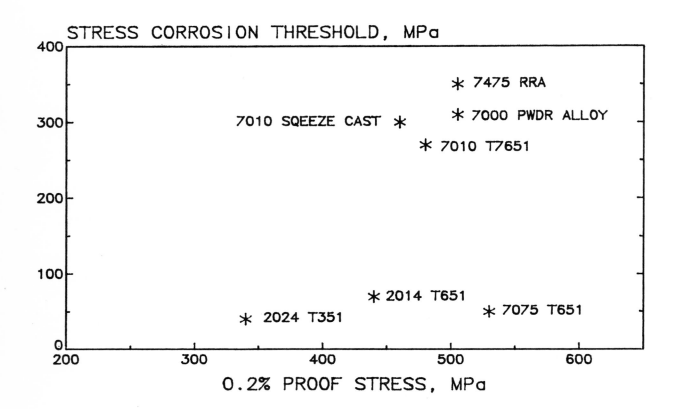

Fig 9. Stress corrosion thresholds for selected conventional alloys demonstrating the marked improvements achieved with special heat treatments.

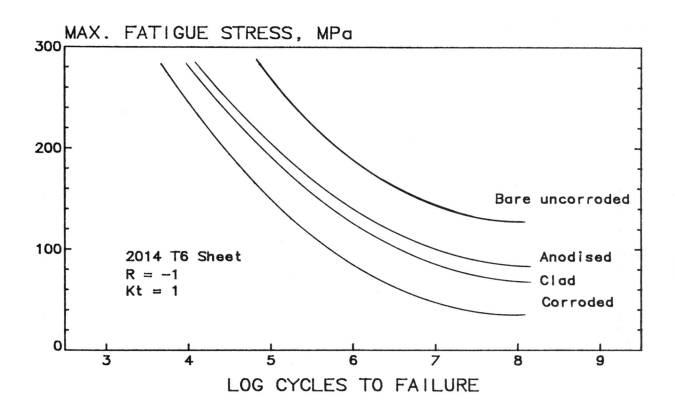

Fig 10. Fatigue performance of conventional
aluminium alloys showing the damaging effects
of cladding and anodising.

of the titanium alloys is very good in this respect.

COMPATIBILITY WITH EXISTING MANUFACTURING TECHNIQUES

Compatibility of any development with existing factory equipment has essentially been considered in the section on cost. It is easy to recognise that a change from aluminium structure to that based upon polymer composite materials requires very major investments and "a new way of life". More subtle limitations may not be so readily recognised. For example, a new alloy that naturally ages more rapidly than conventional material may require cold storage facilities to be improved to allow sheet forming operations to be accomplished without re-heat treatment. A RAPID SOLIDIFICATION RATE material may need to be produced as a close-to-form product because heat treatments used in conventional forging and forming operations destroy the advantageous properties. A change in cold formability may require the introduction of hot forming techniques or superplastic forming with additional expense. The introduction of an acid pickle during the final finishing of aluminium alloy parts may cause problems if no bath is available. Many more examples could be provided.

CONCLUSIONS

This paper has reviewed the limitations imposed upon the use of aluminium alloys in airframe construction and some of the developments that may improve the competitiveness of aluminium-based products. Critical variables were considered to be the availability of large sizes of standard products and the costs of both raw material and manufacture, which critically influence the viability of both existing and new materials. The structural efficiency of aluminium alloys was shown to be mainly deficient in respect to fatigue properties,

an area in which future developments may be able to exploit metal matrix composites or ARALL. Reductions in the basic density of aluminium alloys are soon to be exploited with the lithium containing alloys. It would appear that aluminium alloys will remain the major structural material used in airframe construction this century, but that significant improvements should be implemented if they are to maintain their dominant position in the face of mounting competition from other materials.

ACKNOWLEDGEMENTS

© Copyright Controller HMSO London 1985

REFERENCES

1   W E Quist, G H Narayanan, A L Wingert. 2nd International Aluminium-Lithium Conf, Monterey California AIME. April 1983

2   I F Sakata. Aerospace Congress and Exposition Anaheim California SAE. October 1982

3   C C Bampton, R Raj. Acta Met 30 p2043 1982

4   J W Gunnink, L B Vogelesang, J Schijve. International Congress on Aeronautical Science 2.6.1 p990 1982

5   DOD-NASA. Advanced Composites Design Guide Vol 1-A

6   C J Peel, B Evans, D S McDarmaid. 3rd International Aluminium-Lithium Conf, Oxford, Institute of Metals, July 1985

7   M U Islam, W Wallace. Metals Technology 10 p386 1983

8   R G Mitchell. Corrosion Prevention and Control. June 1981

Aero engine alloy development - the sky's the limit?

D DRIVER

In man's quest for speed and height the sky is almost literally the limit. Such aspirations have, however, been achieved with very expensive cryogenic fuels. For more 'down to earth' air transport fuel more efficient gas turbines are needed. The choice of material to meet the engineering needs of improved propulsive efficiency, lower specific fuel consumption and higher turbine entry temperatures are defined together with the new processing routes which complement the advances made in alloy development.

The author is with Rolls Royce Ltd, Materials Research and Development, Derby.

INTRODUCTION

Material choice is governed by the selection criteria. For materials scientists the choice may be based on such fundamental limitations as melting point, (or strength at temperature), density and chemical reactivity [1] while economists may be restricted by cost and suppliers by manufacturing constraints. For aero-engine development the prime commercial and engineering motivation is to produce fuel efficient engines with low cost of ownership[2]; this paper discusses the resulting material implications.

HISTORICAL CONTEXT

Man first became airborne in 1783, thanks to Montgolfier's hot air balloon, yet it was not for another 120 years that powered sustained flight was achieved. This delay can be attributed to materials/engineering limitations since no engine of sufficient power and low weight was available. Early unsuccessful attempts at powered flight used steam power but in 1902 Charles Manley tested a 5 cylinder petrol driven radial engine weighing less than 200 lbs which developed 52 h.p. Although the Manley engine was used in the unsuccessful Langley "Aerodrome" its attractive power to weight ratio of $\sim$ 0.25 hp : 1 lb. provided the impetus for powered flight. Within a year the Wright brothers achieved the distinction of the first sustained and controlled powered flight with their 4 cylinder water cooled petrol engine which weighed 180 lbs. and developed $\sim$ 16 hp (at $<$ 0.1 hp/lb). The Wright Flyer engine drove two 8 ft diameter propellers to produce a cruising speed of $\sim$ 30 m.p.h. With an estimated propeller efficiency of $\sim$ 66% the aircraft achieved a thrust horse power of . $\sim$ 10.[3] Since those early days subsequent engine development has been spectacular; for example the four Olympus engines, which power Concorde, achieve 150,000 thrust horse power at a cruising speed of Mach 2 ($\sim$ 1520 mph) [3], while the Shuttle spacecraft requires three main propulsion units each developing 375,000 lb thrust plus two solid propellant booster rockets each of 2,900,000 lb thrust to achieve orbital velocities of 18,000 mph. With rockets of 7,500,000 lb thrust sending satellites into outer space at earth escape velocities of 25,000 mph the sky is now literally the limit!

THE COST FACTOR

Although man has aimed for the stars, the cryogenic fuels and oxidants of the rocket motors are also astronomically expensive and specific fuel consumption, (sfc),+ is nearly an order of magnitude greater than for current

---

+ Specific fuel consumption is defined as the amount of fuel consumed (lbs)/hour/lb thrust.

Table II. Nickel – base turbine blade alloys

Alloy	Composition wt %												
	Ni	Cr	Co	Ti	Al	Mo	W	Nb	Ta	C	Zr	B	Others
(a) Wrought													
N75	Bal.	20	–	0.4	–	–							
N80	Bal.	20	–	2.3	1.1	–							
N90	Bal.	20	20	2.4	1.4	–							
N95	Bal.	20	20	2.9	2	–							
N100	Bal.	11	20	1.5	5	5							
N105	Bal.	15	20	1.2	4.5	5							
N115	Bal.	15	15	4	5	3.5							
(b) Cast													
IN100	Bal.	10	15	4.7	5.5	3.0	–	–	–	0.18	0.06	0.014	1.0V
Mar M200	Bal.	9	10	2	5	–	12.5	1.0	–	0.15	0.05	0.015	
Mar M002	Bal.	9	10	1.5	5.5	–	10	–	2.5	0.15	0.05	0.015	1.5Hf

Table III. Single Crystal Alloys

Alloy	Composition wt %								
	Ni	Cr	Co	Ti	Al	Mo	W	Ta	Others
Alloy 444	Bal.	8.6	–	1.98	5.1	–	11.1	–	
Alloy 454 (P&W 1480)	Bal.	10.0	5.0	1.5	5.0	–	4.0	12.0	
SRR 99	Bal.	8.5	5.0	2.2	5.5	–	9.5	2.8	<0.02C
RR 2000	Bal.	10	15	4.0	5.5	3.0	–	–	<0.02C; 1.0V
RR 2060	Bal.	15	5.0	2.0	5.0	2.0	2.0	5.0	<0.02C

Trends in thrust

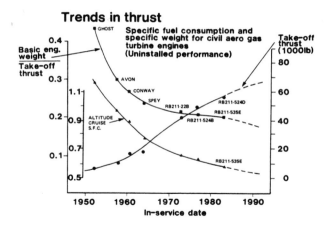

Civil engine cruise S.F.C. trends
Including breakdown of propulsive cycle, thermal cycle and component effects

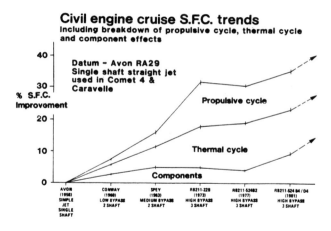

Figure 1. Trends in thrust, sfc and specific weight.

Figure 2. Trends in component, thermal and propulsive efficiencies.

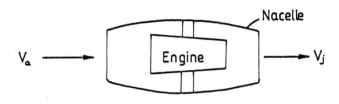

Figure 3. Aero engine propulsive duct.

jet-powered passenger carrying aircraft. Hence for more down to earth trans-continental transportation jet propulsion provides a more cost-effective form of travel, with typical wide-bodied aircraft currently having a fuel consumption which exceeds 80 miles per gallon per person while offering increased speed and hence reduced travelling time over the family car! There is still, however, a strong incentive to reduce fuel costs, particularly since the 1970's when escalating oil prices led to the cost of fuel accounting for more than 50% of the direct operating cost of an aircraft compared with ~ 25% pre-1970.

Figure 1 illustrates how the commercial challenges are being met by dramatic improvements in fuel efficiency and thrust-to-weight ratio with sfc and specific weight * having been

more than halved whilst thrust has increased fifty-fold [4] since the first jet engines were introduced into service.

FUNDAMENTAL LIMITATIONS

Despite the considerable advances made in aero-engine development the gas turbine must conform to the fundamental limitation of all machines, namely that overall efficiency, η_o, can never be greater than unity. This limitation can be written as :-

$$\eta_o = \eta_p \times \eta_T \times \eta_{th} \qquad 1. - - - - (1)$$

where η_p = the propulsive (or Froude) efficiency (and includes installation losses)

---

* Specific weight is the engine weight (lbs)/ engine thrust (lbs).

η_T = the transmission (or transfer) efficiency (and is a function of component efficiency)

and η_{th} = the thermal efficiency.

The overall efficiency, η_o, is also related to the specific fuel consumption by the relationship :- (5)

$$\eta_o = \frac{Va}{sfc} \times \frac{1}{Q \cdot fuel} \qquad ---- (2)$$

where Va = the aircraft speed

and Qfuel = the calorific value of the fuel.

For a 100% efficient engine ($\eta_o = 1$) and a typical calorific value for the aviation fuel of 10,000 CHU/lb. (or $\sim 19 \times 10^6$ Joules/lb or 14×10^6 ft lb/lb) equation (2) reduces to (2):-

$$sfc_{ideal} \approx \frac{Ma}{4} \quad lb \; fuel/lb \; thrust/hr$$

where Ma = the aircrafts Mach number (Mach 1 = 760 mph)

Hence for a typical aircraft cruising speed of Mach 0.8 the theoretical limiting sfc (lb/hr/lbf) is 0.2. Table I gives the breakdown of losses in a typical large turbofan engine, such as the RB 211, from which it can be seen that only about one third of the fuel consumed is used to propel the aircraft.

TABLE I

Breakdown of Losses in a Modern Gas Turbine Engine.

Source of Loss		Sfc lb/hr/ lbf	η_o %
Fully efficient engine		0.20	100
Propulsive/installation losses	η_p	+0.19	−12
Transfer/component losses	η_T	+0.18	−27
Thermal losses	η_{th}	+0.09	−31
Current turbofan efficiency		0.66	30

Figure 2 illustrates historically how improvements to sfc have been made. Until the mid-1960's improvements in all aspects of sfc were steady but the introduction of the high bypass ratio RB 211 engine in the early 1970's with its appreciably higher mass flow resulted in a significant increase in propulsive efficiency, while improvements in materials and blade

cooling produced greatly increased turbine entry temperatures (TET) with corresponding improvements in the thermal cycle efficiency. Until the late 1970's few improvements were made in component efficiencies but powerful computing facilities now allow detailed air flow calculations and three dimensional aerodynamic component designs to be carried out relatively quickly and the resulting improvements in component efficiencies have been dramatic. In the remainder of this article we discuss the engineering and materials interactions which are enabling aero-engine development to approach more closely the fundamental limitations in efficiency.

LIMITATIONS IN PROPULSIVE EFFICIENCY

The propulsive efficiency, η_p, is defined as the ratio of useful energy used to propel the aircraft compared with the total energy of the jet. This latter term includes not only the propulsive energy but also the kinetic energy imparted to the jet, (Fig 3), such that for a **given** engine thrust :-

$$F = m \, (Vj - Va) \qquad ---- (3)$$

where m is the mass flow through the engine (assumed constant) and Vj and Va are the jet and aircraft velocity respectively, then:- (2.5)

$$\eta_p = \frac{2Va}{Va+Vj} = \frac{2}{1 + (Vj/Va)} \qquad ---- (4)$$

Maximum propulsive efficiency ($\eta_p = 1$) would be achieved if Vj = Va) but this would be meaningless since the aircraft would have no thrust, (F = 0). The aim is therefore to achieve sufficient thrust for propulsion within the chosen speed range whilst maintaining only a small difference between Vj and Va at cruise in order to maximise propulsive efficiency. Since the necessary thrust can be achieved, (equation 3), by high mass flow at low velocity (as with a propeller) or by low mass flow with high velocity, (as with a turbojet), a range of propulsive units can be developed with different mechanical configurations but optimum propulsive efficiency within a given speed range. Figure 4 illustrates engine configurations in which the propulsive efficiency is optimised for progressively higher speeds by decreasing the mass flow and increasing the jet velocity.

* [In recent years the enthusiasm for relatively high fuel burn supersonic travel has been tempered by the cost advantages of subsonic flight and greater attention has therefore been given to optimised engine configurations in the 0.5 − 0.9 Mach speed regime. At low speeds, (up to 0,5 Ma; 400 mph) propeller driven aircraft have the greatest propulsive efficiency but as

---

* The bracketed section has been included in response to questioning at the meeting.

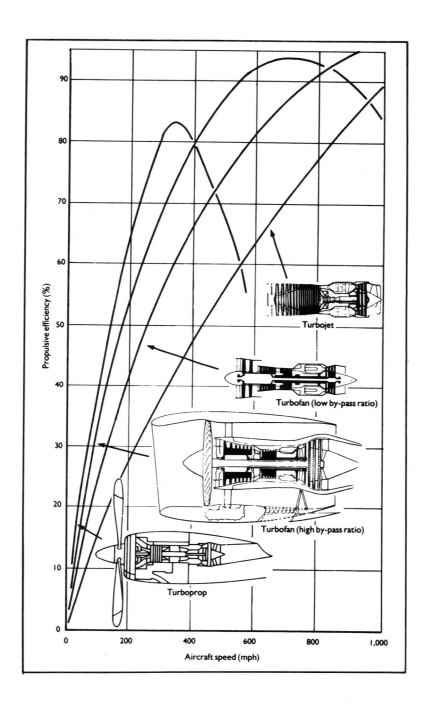

Figure 4.   Propulsive efficiency of various
engine types.

flight speeds increase to between 400 and 800 mph (0.5 - 1 Mach) the turbofan predominates, (Fig.4). Increased mass flow through the turbofan engine could be achieved by increasing the by-pass ratio.† Unfortunately the increased drag of the larger diameter fan cowling offsets the advantage of higher mass flow, thereby making bypass ratios of greater than 6 : 1 impractical. Alternatively the fan cowling can be eliminated thereby converting the turbofan into a large propeller driven aircraft. The reason why large diameter multi-bladed propellers have not been used earlier is because as rotational speeds increase, (to achieve high mass flow), propeller tip speeds become supersonic and aerofoil air flow becomes turbulent. By twisting the propeller blade such that the tip sections present a shallower angle to the direction of flight then turbulence can be reduced.

Experiments by Hamilton Standard and NASA on swept back propellers with very thin aerofoil sections, (Fig.5), have recently demonstrated that such "propfans" can achieve significantly improved propulsive efficiencies compared with conventional propellers especially when a second contra-rotating set of blades is added to recover swirl losses. Figure 6 illustrates the propfan propulsive efficiency relative to conventional turboprops and turbofans, while Figure 7 shows one of the geared contra-rotating propfan arrangements. In order to minimise noise and vibration the propfan should preferably be mounted at the rear of the aircraft with the propellers behind the engine powerplant in a "pusher" configuration (Fig.7). In consequence hot gases will pass through the propeller hubs thereby necessitating a light, strong, but relatively high temperature material such as titanium.]

LIMITATIONS IN TRANSFER EFFICIENCY

The ability of an aero engine to compress the incoming air and to extract energy from the gas in the turbine assembly is critically dependent on the efficiency of the individual components, (ie compressor and turbine blades), which are controlling the energy transfer. For bypass engines, such as the RB 211 (Fig.8), it also depends on what proportion of gas is transferred to the bypass duct. An approximate relationship for the transfer efficiency, η_T, is given by[2]:-

$$\eta_T = \frac{\mu\,\eta_F\,\eta_t + 1}{\mu + 1} \quad ----\ (5)$$

where μ is the bypass ratio, (which for current civil high by-pass engines, such as the CF6, JT9D and RB 211 is typically $\sim$ 5 : 1),

while η_F and η_t are the fan and turbine component efficiency.

In order to keep the overall length of the engine as short as possible, each compressor stage needs to achieve a high pressure ratio. Figure 9 illustrates how fan blade efficiencies above 90% have facilitated by-pass pressure ratios approaching 2 : 1, one obvious advantage in aerodynamic efficiency and reduced weight being the transition from the 33 bladed, solid, clappered titanium fan of the RB 211-22B (Fig.10), to the 22 bladed, hollow, wide chord fan assembly on the RB 211-535C (Fig.11).

MATERIAL LIMITS FOR FAN BLADES

To achieve a relatively high mass flow through the engine, the compressor blades need to rotate quickly. The resulting centrifugal tensile stresses, σ_{cf}, which are a maximum at the blade root, are given by[5]:-

$$\sigma_{cf} = \frac{\rho_b\,w^2}{ar}\int_r^t a.r\,dr \quad ----\ (6)$$

where ρ_b is the density of the blade material
  w is the angular velocity of the blade
and ar is the cross sectional area of the blade at the root.

In modern blades the cross sectional area, a, and the aerofoil radius of curvature, r, vary from root (r) to tip (t).

As the blade length and rotational speed increase there is a corresponding increase in centrifugal loading. In consequence preference is given to compressor blade materials with a high specific strength, (ie strength/density), over the required temperature range. At 30,000 feet temperatures fall to -50°C while in the hottest climates ambient temperatures only occasionally exceed +50°C. This defines the temperature range over which the fan blades might be expected to operate. A comparison of the specific tensile strength of potential compressor blade materials, (Fig.12) demonstrates the attractions of glass reinforced plastics, (GRP), and carbon fibre composites for this application with titanium predominant for the later stages of the compressor where temperatures rise between 200 and 600°C.

Although high aspect fan blades are efficient, they are prone to vibration from mechanical excitation or aerodynamic flutter. Such vibrations are similar to those in a uniform cantilever beam for which the resonant frequency, f, is given by[6]:-

$$f = \frac{K}{\ell^2}\sqrt{\frac{EI}{\rho}} \quad ----\ (7)$$

---

† The by-pass ratio is the ratio of (cool) air flow which is directed by the fan blades around the core of the engine and later combined with the exhaust stream, to that (hot) air flow which goes through the core.

UNI-ROTATION TRACTOR

Figure 5.   Thin supersonic swept-back 'propfan'.

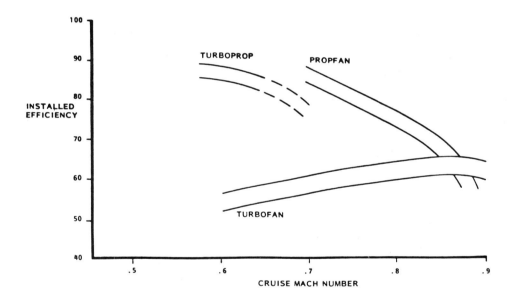

Figure 6.   Comparison of sub-sonic propulsive efficiencies.

Advanced contraprop - integral gearbox and hub

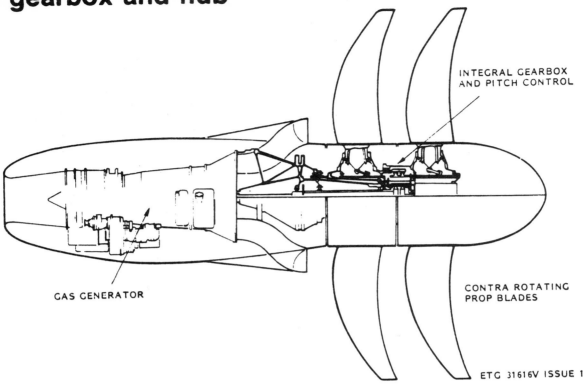

INTEGRAL GEARBOX
AND PITCH CONTROL

GAS GENERATOR

CONTRA ROTATING
PROP BLADES

ETG 31616V ISSUE 1

Figure 7.   A typical contra-rotating geared
             propfan.

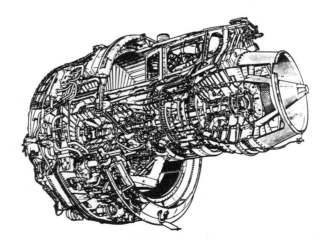

Figure 8.   Cut-away of an RB211 aero-engine.

70

Single stage fan progress

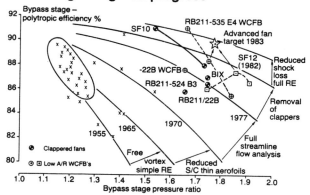

Figure 9.   Single stage fan progress.

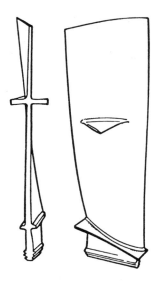

Snubbered fan blade

Figure 10.   A clappered fan blade assembly.

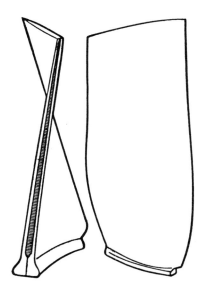

Wide-chord fan blade

Figure 11.   Wide chord fan blades.

where K is a constant; ℓ is the length of the component;
E is Youngs modulus; I is the moment of inertia
and ρ is the density

Materials with high specific modulus, (E/ρ), help push the blades resonant frequency outside the engines operating range. Figures 12 and 13 illustrate that for both high specific strength and stiffness the carbon fibre fan blade offers particular advantages even when it is recognised that these properties are governed by the volume fraction and alignment of the fibres and that the values illustrated are in the optimum strength direction, (ie parallel to the aligned fibres).

The attractions of carbon fibre fan blades were sufficient to encourage a strong manufacturing commitment in the mid 1960's [7] following successful demonstration of the concept by operating trial sets of carbon-fibre Conway LP compressor blades for some 4,000 hours in service[8]. Unfortunately, the larger scale and more complicated geometry of the RB 211 fan blade presented more difficult manufacturing problems than with the Conway blade; in particular blade damage from high speed impact and erosion caused delamination which required three dimensional reinforcement, while root fixing and effective bird strike capabilities were additional difficulties. In view of the limited development time, the carbon fibre fan blade was withdrawn and replaced by the solid titanium fan blade, although the experience gained in carbon fibre technology has subsequently been used to advantage in the manufacture of RB 211 engine nacelle and thrust-reverser components[8].

The ability to forge the relatively isotropic low density titanium alloy Ti-6Al-4V into near nett shape motivated its choice as a fan blade material for the JT9D, CF6 and RB 211 engines, though the large size of these components precluded precision forging and necessitated expensive machining of the oversize forging. More recent developments have now enabled such components to be precision forged. The relatively poor specific stiffness of titanium alloys, (Fig.13), has however required that the blades incorporate geometric stiffening features, (termed "snubbers" or "clappers"), at approximately blade mid-height such that adjacent snubbers abut against each other thereby providing a supporting ring against lateral movement, (Fig.10). Although this form of increased rigidity has been introduced on all early high bypass engines, the solution is hardly elegant and imposes a 1-2% loss in sfc due to aerodynamic losses and increased weight.

When faced with the need for lightweight stiff structures the packaging industry came up with a far more satisfying solution with the introduction of corrugated cardboard in which the laterally stiffening corrugations are sandwiched between two thin sheets! A similar principle was developed within Rolls-Royce for the wide chord fan blade which involved vacuum brazing a titanium honeycombe core between two sheets of titanium which were creep formed and chemically etched to generate the correct aerofoil shape, (Fig.14). Such wide chord fan blades are now in service on the 74 in. dia. fan assembly of the RB 211-535 E4.

As with so many aero-engine developments it is the manufacturing methods rather than the material characteristics which tend to limit the full exploitation of both current and new materials. The process capabilities for future materials will, quite literally, govern the shape of things to come and to this end superplastic forming and diffusion bonding of titanium alloys have been demonstrated for possible future fan blade constructions. Improvements in high temperature materials for both turbine disc and blade applications are similarly process dominated.

LIMITATIONS IN THERMAL EFFICIENCY

For an ideal (Joule or Brayton) cycle of a simple gas turbine, (Fig.15), the thermal efficiency of the cycle, η_{th}, is given by (5):-

$$\eta_{th} = \frac{\text{net work output}}{\text{heat input}} = \frac{Cp(T_3-T_4)-Cp(T_2-T_1)}{Cp(T_3-T_2)} \quad ----(8)$$

Making use of the relationships[5]:-

$$\frac{P2}{P1} = \frac{P3}{P4} = r \text{ (pressure ratio) and } \frac{T2}{T1} = \frac{T3}{T4} = r^{\frac{(\gamma-1)}{\gamma}}$$

where $\gamma = Cp/Cv$ where Cp = specific heat at constant pressure
and Cv = specific heat at constant volume

then equation (8) reduces to the simpler form*:-

$$\eta_{th} = 1 - \frac{T_4}{T_3} = 1 - \left(\frac{1}{r}\right)^{\frac{\gamma-1}{\gamma}} \quad ----(9)$$

When component losses are taken into account the apparently independent relationships of temperature and pressure ratio with thermal efficiency become combined as shown in Figure 16. Alternatively the variation in gas turbine performance with pressure ratio and turbine entry temperature, (TET), can be expressed as a function of sfc and specific thrust as in Figure 17. Both figures demonstrate the same trend, namely that for maximum fuel efficiency or high thrust with minimum sfc aero engines need to operate with high compression ratios and high TET.

Early jet engines, (such as the Whittle W1), had a compression ratio of 4 : 1 and a TET of 970°K. This contrasts with modern aero-engines, such as the RB 211, where compression ratios of 30 : 1 are achieved and TET's approach 1700°K while still higher temperatures have been demonstrated on rig testing (Fig.18). The ability to operate turbine blades at progressively higher temperatures has been due to both material advances and to the considerable developments made in blade cooling, (Fig.19), the relative contribution from each source being illustrated in Figures 18 and 20.

Typically the average increase in turbine blade temperature capability due to material improvements has been somewhat greater than 10°K per annum while progress from blade cooling has averaged 30°K per annum.

BLADE COOLING LIMITATIONS

Heat is extracted from turbine blades by forced convection, the cooling air being pumped through longitudinal passages, (Fig.21). In their simplest form the passages are circular, elliptical or triangular, (Fig.19 up to 1975), such

---

* Note the typographical errors in reference 1.

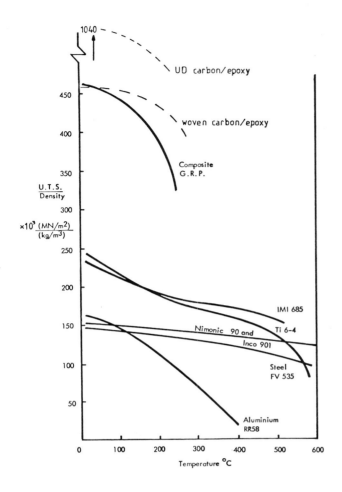

Figure 12.   Specific tensile strength of
compressor materials.

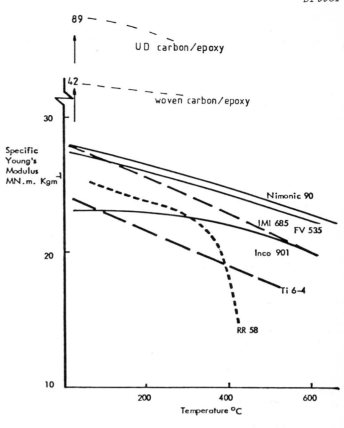

Figure 13.   Specific modulus of compressor
materials.

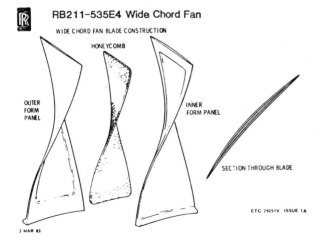

Figure 14.   Manufacturing process for wide
chord fan blades.

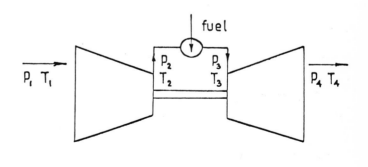

Figure 15.   Ideal cycle for a simple gas
turbine.

Thermal efficiency

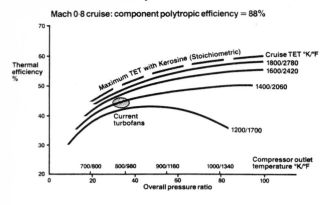

Figure 16. Variation of thermal efficiency with pressure ratio.

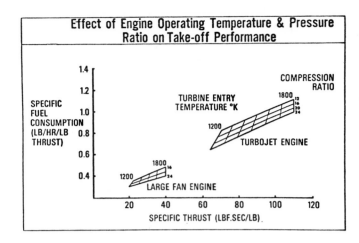

Figure 17. Effect of turbine entry temperature and compression ratio on specific thrust and sfc.

Trend in turbine gas temperature

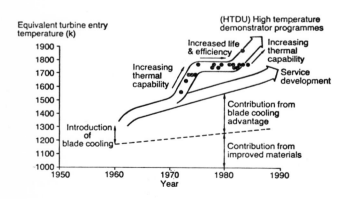

Figure 18. Trends in turbine entry temperatures.

pipe flow cooling being generated by drilling or by the etching out of metal from drilled and plugged holes following forging or extrusion[10]. Mould assemblies with silica tube insertions provide a further method of introducing cooling passages into cast components. Examples of this relatively simple form of turbine blade cooling were found on the Rolls-Royce Conway, Spey and Olympus engines together with early American engines such as the Pratt and Whitney TF30 and Allison T56,[11] this last engine having the distinction of being the first American engine to enter service with cooled turbine blades, (in September 1964).

More complex internal surfaces involve radial fins, pedestals and cross flow fins, (Fig.22), these extended surfaces being introduced to increase the heat flux between the metal and the cooling air and to provide obstacles to the coolant flow, thereby increasing turbulence and raising the heat transfer coefficient.[12] For such complicated internal cooling configurations precision investment casting by the lost wax process, using complex cores, is one of the few manufacturing methods.[13]

An example of a cast RB 211 high pressure turbine blade is shown in Figure 23 with the aerofoil surface cut away to reveal the internal cooling geometry, while the three dimensional illustrations of Figures 24 and 25 show more graphically the complexity of current turbine blade and nozzle guide vane cooling.

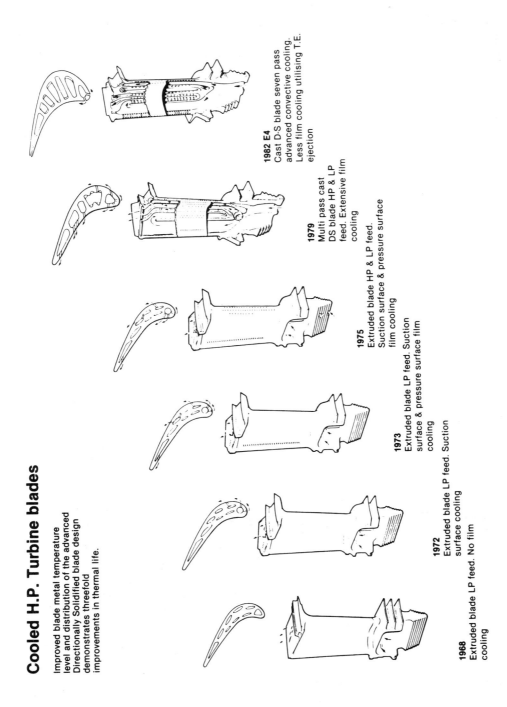

Cooled H.P. Turbine blades

Improved blade metal temperature
level and distribution of the advanced
Directionally Solidified blade design
demonstrates threefold
improvements in thermal life.

1982 E4
Cast D-S blade seven pass
advanced convective cooling.
Less film cooling utilising T.E.
ejection

1979
Multi pass cast
DS blade HP & LP
feed. Extensive film
cooling

1975
Extruded blade HP & LP feed.
Suction surface & pressure surface
film cooling

1973
Extruded blade LP feed. Suction
surface & pressure surface film
cooling

1972
Extruded blade LP feed. Suction
surface cooling

1968
Extruded blade LP feed. No film
cooling

Figure 19.   Trends in turbine blade cooling.

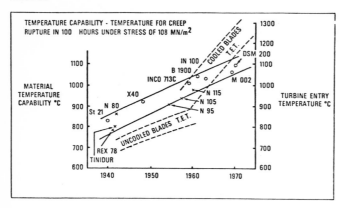

Figure 20. Contributions from Materials and blade cooling to increases in TET.

Figure 21. Cooling configurations in Conway (left) and RB211 (right) high pressure turbine blades.

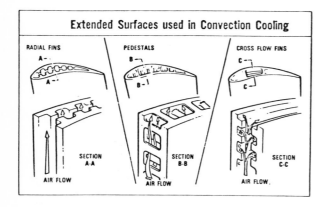

Figure 22. Cooling configurations in turbine blades.

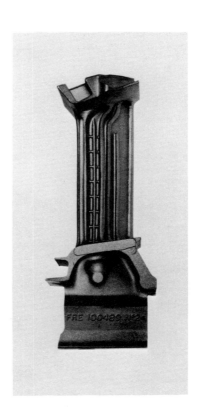

Figure 23. RB211 high pressure turbine blade with aerofoil surface removed to reveal internal cooling passages.

535E4 HP turbine blade

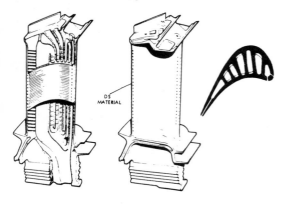

Figure 24. Illustration of cooling configurations in a complex cooled turbine blade.

Since the cooling air in the passages possesses a large residual thermal capacity as it leaves the blade, the air can be used to form an additional protective cooling film on the aerofoil surface. Such film cooling is economically achieved by cutting fine holes through the aerofoil surface into the underlying cooling passages using such techniques as electro-discharge machining or electron beam or laser drilling, the combination of convective and film cooling being sufficient to enable modern blade materials to withstand $2000^{o}K$ gas temperatures.

For gas temperatures greater than $2000^{o}K$, transpiration cooling will be necessary whereby air cooling is forced through a porous wall. A transpiration cooling system constitutes the theoretical cooling limit since it includes an almost 100% efficient convective system prior to coolant ejection through the pores followed by the formation of a continuous protective gas film over the surface [12]. However, material and manufacturing problems, together with performance penalties, have tended to preclude this form of turbine blade cooling, though a pseudo-transpiration cooling technology (termed transply) has been developed by Rolls-Royce for sheet combustor applications.

MATERIAL LIMITS FOR TURBINE BLADES

Advances in blade cooling technology have been matched by associated materials and manufacturing development such that by the 1970's the limited high temperature strength capabilities of the wrought nickel base superalloys were being superseded by cast alloys, originally in polycrystalline form and subsequently as directionally solidified and single crystal components,[14.15] (Fig.26). Figure 26 also illustrates how the progressive increases achieved by alloy development can be complemented by step changes brought about by new processing technology. Early wrought alloys developed from the simple 80Ni-20Cr Nimonic 75 through to N115 by gradually increasing the volume fraction of the γ' (Ni_3Al) phase through increased (Ti + Al) additions coupled with solid solution strengthening of the matrix by cobalt and molybdenum additions, (Fig.27). The introduction of vacuum casting produced a significant increase in temperature capability since higher (Ti + Al) contents could now be retained in the melt without loss by oxidation, giving γ' volume fractions of up to 70% in such alloys as IN100, (alloy compositions are given in Table II). It is impractical to increase (Ti + Al) contents beyond those in IN100 since little more of these elements can be taken into solution and precipitated as γ'. Instead increased matrix strength has been developed by the addition of high melting point elements with a relatively large atomic size, such as tungsten, typical examples of this family being Mar M200, a turbine blade alloy favoured by Pratt and Whitney and Mar M002 used by Rolls-Royce.

A particular characteristic of the Martin Metals, (Mar M--), series of nickel-base alloys is the patented addition of hafnium [16-18], some 2 wt.% of this element conferring improved ductility by modifying the solidification sequence. In particular hafnium is rejected from the solidifying dendrites during casting and concentrates in the interdendritic regions thereby increasing the quantity and coarsening the form of the γ - γ' eutectic. Hafnium also modifies the size, distribution and composition of the grain boundary carbides, producing a finer dispersion.

A further step change, (Fig.26) accompanied the transition from conventional casting to directional solidification.[14-15]. Instead of allowing the molten metal within a mould to solidify by natural heat losses thereby producing a casting with a conventional "chill" outer structure, an intermediate columnar region and an equiaxed central core, directional solidification imposes a controlled thermal gradient on the solidifying metal. In the system illustrated in Figure 28 the mould is enclosed within the hot zone of a furnace from which the majority of heat is removed through a water-cooled base plate.[19-20]. Nucleation of the metal within the mould therefore proceeds from the chill. As the mould is withdrawn from the furnace columnar growth proceeds from the existing nuclei generating directionally solidified components in which transverse grain boundaries have been eliminated, the remaining grain boundaries lying parallel to the blade axis (Fig.29). Since intergranular creep failure on transverse grain boundaries, which are perpendicular to the applied centrifugal loading, is a common failure mode for turbine blades, the elimination of these grain boundary sites increases the life of the component. Geometric constrictions can be incorporated into the mould, (Fig.30), through which only one of the growing columnar grains can pass thereby generating a single crystal, (Fig.31). Since the favoured growth direction for face-centred cubic nickel base alloys is $\langle 100 \rangle$ this orientation is aligned along the blade axis providing both high creep resistance and good thermal fatigue, the latter characteristic being due to the low Youngs modulus in the $\langle 100 \rangle$ direction which enables thermal stresses to be minimised. If full three dimensional crystallographic control is to be established, the single crystals need to be grown from suitably oriented seed crystals located in the base of the mould prior to casting. Close proximity of the seed to the chill plate, coupled with the preferred side entry, bottom running system of Figure 32, ensures consistent melt-back of the seed crystal with subsequent epitaxial growth.[19.20].

In order to increase component life and reduce the possibility of stray grain nucleation from oxide entrapment, a clean melt is essential. During melting, surface oxide and refractory pick-up can form a slag on top of the melt which, if then poured into the mould assembly will generate deleterious inclusions. The use of an induction melting, bottom pouring crucible system,

77

Advanced 3D nozzle guide vane

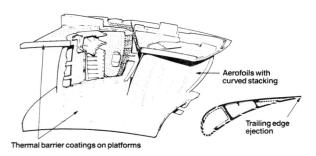

Aerofoils with curved stacking

Trailing edge ejection

Thermal barrier coatings on platforms

Figure 25. Illustration of cooling configurations in a nozzle guide vane.

Turbine blade alloy capabilities

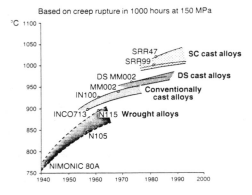

Based on creep rupture in 1000 hours at 150 MPa

Figure 26. Temperature capability of turbine
blade alloys.

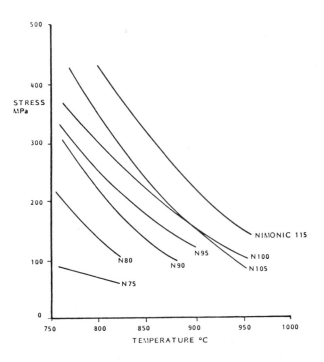

Figure 27. High temperature properties of
Nimonic alloys.

The Rolls-Royce DS Furnace

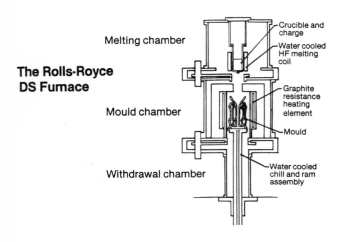

Melting chamber

Crucible and charge

Water cooled HF melting coil

Graphite resistance heating element

Mould chamber

Mould

Withdrawal chamber

Water cooled chill and ram assembly

Figure 28. Main features of Rolls-Royce
directional solidification plant.

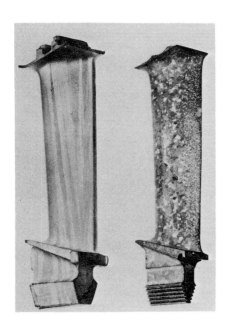

Figure 29. Comparison between a directionally
solidified (left) and convention-
ally cast (right) turbine blade.

(Figs.28 and 33), helps eliminate these difficulties and provides consistent melting time and temperature control without recourse to extensive monitoring and feed-back systems[20]. A thin alloy disc, ("penny"), is placed over the hole in the bottom of the crucible and the metal charge added, (Stage 1, Fig.33). By suitable arrangement of the induction melting field, the top of the charge can be melted first, (Stage 2 Fig.33) and the magnetic 'pinch' effect of the high frequency induction melting creates essentially levitation melting with very little contact, (and hence refractory pick-up), with the crucible (Stage 3, Fig.33). The degree of melt superheat is controlled by the thickness of the penny which is melted last by conduction to give the automatic bottom pouring, (Stage 4, Fig.33). Most oxide from the original charge will float to the surface of the melt as a low density slag and be retained within the crucible, (Fig.34), though additional ceramic filtering can also be introduced either in the mould ingate or immediately below the main body of the casting, (Fig.32) to ensure a high level of cleanliness[20].

With their reduced grain boundary area, there exists scope for developing alloys specifically for directional solidification [21-23], though it has been more common practice for aero-engine manufacturers to process existing cast alloys via a directionally solidified route. It is in the development of alloys for single crystal technology that most attention has been directed. The elimination of grain boundaries facilitates the removal of those elements which were primarily introduced for their grain boundary strengthening role, namely carbon, zirconium and hafnium. Since these elements tend to lower the incipient melting point, their removal allows higher solution heat treatment temperatures with a corresponding improvement in chemical homogeneity and a more uniform γ ' distribution. The simplified chemistry also allows some adjustment of the γ ' forming elements without producing metallurgical instability though care still needs to be maintained in order to ensure a sufficient gap between the incipient melting point and the γ ' solvus to enable the alloy to be effectively heat treated. In some alloys, where the heat treatment window, (HTW), is small, several increasingly higher heat treatment sequences are used to progressively homogenise the structure.

In some circumstances the processing characteristics of the directional solidification furnace can influence the scope for alloy development. For example, some directional solidification plants have a shallow temperature gradient which produces an extended 'mushy' zone ahead of the solidifying interface. In such circumstances high tungsten bearing alloys can produce a characteristic processing defect termed 'freckling', which consists of a stringer of small equiaxed grains within the body of an otherwise single crystal.[24.25] Freckling is caused by preferential partitioning of dense elements to the solidifying dendrites, while the lighter γ ' forming elements titanium and aluminium are rejected into the liquid. The resulting density inversion provides a concentration-driven convection current which erodes the solidifying dendrite tips and the debris provides nuclei for the randomly oriented freckle grains. If the alloy chemistry can be adjusted so that dense elements are also rejected into the melt to compensate for the titanium and aluminium rejection then freckle formation will be minimised. This philosophy was adopted by Pratt and Whitney when developing their single crystal alloy 454, (PW 1480). Tungsten partitions equally between the solidifying dendrites and the melt whereas the equally dense tantalum is rejected almost exclusively into the melt. Partial replacement of tungsten by tantalum thereby minimises freckling, (compare alloy 444 with 454 in Table III).

The Rolls-Royce directional solidification plant possesses a steep thermal gradient with a corresponding narrow 'mushy' zone, thereby minimising the tendency for freckling and providing greater opportunities for alloy development, (Fig.35). With this greater freedom, a family of single crystal alloys has been developed to meet the individual requirements of particular aero-engine components[26]:-

  (i) SRR 99 - a high strength, high temperature alloy containing both solid solution strengthening elements and a high volume fraction of γ ' precipitate.

 (ii) RR 2000 - a low density alloy based on IN100 which allows blades to operate under high rotational speeds without excessive centrifugal loadings

and (iii) RR 2060 - a nozzle guide vane alloy with increased chromium additions for good hot corrosion resistance in the uncoated condition.

Further single crystal alloys could be developed but as the temperature capability of these nickel-base alloys approaches ever closer to their fundamental limitation - the melting point of the alloy - it is questionable whether additional major commitment is justified. Already single crystal turbine blade alloys are operating at temperatures closer to their melting point than almost any other structural engineering material and the search must therefore continue for new families of materials for the next 'quantum jump' in temperature capability.

Much attention is currently being given to the exploitation of ceramics in monolithic, composite and coated form for future aero-engine applications, provided their intrinsically limited ductility can be allowed for in the design. Some engine running experience has been gained at Rolls-Royce with ceramic shroud rings, bearing

Single crystal selector designs

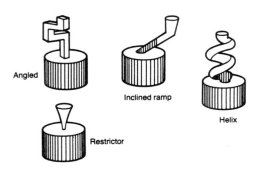

Angled

Inclined ramp

Helix

Restrictor

Figure 30.   Single crystal selector designs.

Single crystal mould design

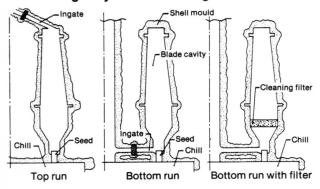

Ingate

Shell mould

Blade cavity

Cleaning filter

Chill   Seed

Ingate   Seed   Chill

Chill

Top run   Bottom run   Bottom run with filter

Figure 32.   Mould designs for single crystal castings.

4-Stage charge melting

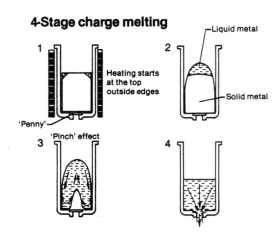

1   Heating starts at the top outside edges

'Penny'

2   Liquid metal

Solid metal

3   'Pinch' effect

4

Figure 33.   Initial charge melting sequence for directional solidification and single crystals.

Figure 31.   A single crystal blade embodying a directionally solidified starter block and helical constriction.

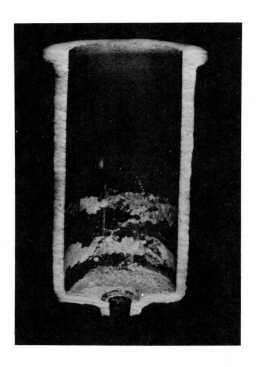

Figure 34.   Dross remaining in crucible after melting cycle.

Alloy development philosophy

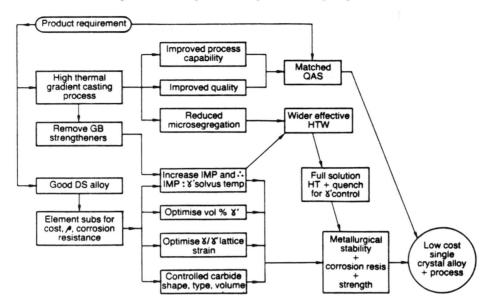

Figure 35.   Possible network for single
crystal alloy development.

components and turbine blades using a Gem
helicopter engine as the test vehicle.   The use
of ceramics for major rotating parts before the
turn of the century has however been questioned
(27).

Ultimately we are limited only by the skill and
imagination of materials scientists, manufacturers and engineers in developing new processing
and behavioural models to meet our engineering
needs.   The sky's the limit!

REFERENCES

1. Driver, D.   Metals & Materials (1985) 1
        357

2. Bennett, H.W. Proc. Inst. Mech. Engrs.
        (1983) 197A. 149

3. Hooker, S.G. J Roy Aer. Soc. (1970) 74   1

4. Meetham, G.W. Metallurgist & Mat. Tech.
        (1976) 11 550

5. Cohen, H., Rogers, G.F.C. & Saravanamutto,
        H.I.H., "Gas Turbine Theory" Pub.
        Longman (1972) 2nd Ed p210

6. Ryder, G.H., Strength of Materials (3rd Ed)
        Cleaver – Hume (1961) p.310

7. Gresham, H.E. Metals & Materials Nov. (1969)
        p.433

8. Turner, W.N., Johnson J.W. & Hannah, C.G.,
        Chapter 5 ref.9

9. Meetham, G.W., (Ed) The Development of Gas Turbine Materials Appl. Sci. Pub (1981)

10. Wells, C., Chapter 8 of ref.9.

11. Halls, G.A., AGARD presentation 18th May (1967) Varenna Italy

12. Fullagar, K.P.L., Presentation at Symposium on the Design & Calculation of Constructions Subject to High Temperatures -- University of Tech. Delft Netherlands 19th Sept. (1973)

13. Ford, D.A., Chapter 6 of ref.9.

14. Alexander, J.D., Proc. Inst. Mech. Engrs (1983) <u>197</u> 75

15. Driver, D. Metals Forum (1984) <u>7</u> 146

16. Duhl, D.N. and Sullivan, C.P., J Metals (July 1971) 38

17. Doherty, J.E., Kear, B.H. and Giamei A.F., J Metals Nov. (1971) 59

18. Kotval, P.S., Venables, J.D. and Calder, R.W Met. Trans (1972) <u>3</u> 453

19. Goulette, M.J., Spilling, P. and Arthey R.P "Superalloys 84", Eds; Gell, M., Kortovich, C.S., Bricknell, R.H., Kent, W.B., Radavich, J.F., Met. Soc. A.I.M.E. (1984) p.167

20. Higginbotham, G.J.S., Mat. Sci. Tech (1986) - to be published

21. Harris, K., Erickson, G.L., and Schwer, R.E. 5th Int. Symp. on Superalloys. Champion P.A. (1984)

22. Erickson, G.L., Harris, K. and Schwer, R.E. ORL/AIME Superalloy Conf. Bethesda M.D. (1984)

23. Erickson, G.L., Harris, K., and Schwer, R.E. Gas Turbine Conf. Houston Texas March (1985)

24. Giamei, A.F. and Kear, B. Met Trans. (1970) <u>1</u> 2185

25. Copley, S.M., Giamei, A.F., Johnson, S.M. and Hornbecker, M.F., Met. Trans. (1970) <u>1</u> 2193

26. Ford, D.A. and Arthey, R.P., "Superalloys 84" Met. Soc. AIME (1984) p.115

27. Meetham, G.W., Metal Bulletin Conf. on Ti & Superalloys N.Y. (1984)

Book 392
Published in 1986 by

The Institute of Metals
1 Carlton House Terrace
London SW1Y 5DB

and

The Institute of Metals
N American Publications Center
Old Post Road
Brookfield
Vt 05036
USA

British Library Cataloguing in Publication Data

Materials at their limits: proceedings of
   the conference held as part of the first
   autumn meeting of The Institute of Metals
   on 25 September 1985 at the University of
   Birmingham.
   1. Materials
   I. Institute of Metals (1985)
   620.1'1   TA403

   ISBN 0-904357-86-4

Library of Congress Cataloging in Publication Data

Materials at their limits

   1. Alloys-Congresses.   2. Ceramics--Congresses.
   3. Composite materials--Congresses
   I. Inst. of Metals. Autumn meeting (1st : 1985 :
London, England)
   TA483.M37 1986   620.1'1   86-7301
   ISBN 0-904357-86-4

Compiled by the Institute's CRC Unit from
typescripts and illustrations provided by
the authors.

Printed and made in England by
Adlard & Son Ltd, Dorking, Surrey

#13424600

mL

MATERIALS AT THEIR LIMITS

Proceedings of the Conference
held as part of the first
Autumn Meeting of
The Institute of Metals
on 25 September 1985
at The University of Birmingham

D
620.11
MAT

The Institute of Metals

London

1986